NIST Technical Note 1608

Building Networks and Public Safety Communications

David G. Holmberg
Stephen J. Treado
Steven T. Bushby
Michael Galler
Kent A. Reed
Alan Vinh
Building Environment Division

William D. Davis
Robert Vettori
Fire Science Division

*Building and Fire Research Laboratory
National Institute of Standard and Technology
Gaithersburg, MD 20899-8530*

October 2008

U.S. DEPARTMENT OF COMMERCE
Carlos M. Gutierrez, Secretary

NATIONAL INSTITUTE OF STANDARDS AND TECHNOLOGY
Patrick Gallagher, Acting Director

Abstract

The Building Networks and Public Safety Communications project has three components: (1) addressing feasibility of using building networks for bridging first responder radio communications into buildings, (2) recommending changes to the SAFECOM Statement of Requirements that address the role of the building, and (3) a workshop that addresses current "what works" for in-building radio communications.

Radio coverage for emergency response in buildings is frequently problematic due to the shielding effects of building structures. Since many public safety incidents occur in buildings, use of existing building infrastructures to bridge radio communications is desirable. This report presents research exploring the potential for using building networks as a communication bridge for radio communications. Results show that building information technology (IT) and fire alarm system networks could potentially be used for routing emergency communications with some changes, while distributed antenna systems are the current preferred solution. Simulations of emergency responder voice, video, and data communications routed across building networks is shown to be feasible using Institute of Electrical and Electronics Engineers (IEEE) 802.11g access points.

In addition to addressing requirements and feasibility of bridging radio communications, this report addresses requirements for a building data interface to public safety networks, and progress towards enabling public safety user access to building data. Emergency response scenarios prove useful for interface requirements development. Recommended changes to the Statement of Requirements are provided along with results from the High-Rise and Large/Complex Incident Communications Workshop.

Keywords: building incident; communications; distributed antenna system; emergency response; in-building wireless; interoperability; networks; public safety; radio; simulation

Disclaimer

Certain trade names or company products are mentioned in this document to describe adequately the work that was performed, give examples of trends in industry, and name software tools used in performing research. In no case does such identification imply recommendation or endorsement by the National Institute of Standards and Technology, nor does it imply that the software and/or hardware used is the best available for the purpose.

In addition, we have provided links to websites that may have information of interest to our users. NIST does not necessarily endorse the views expressed or the facts presented on these sites. Further, NIST does not endorse any commercial products that may be advertised or available on these sites.

Table of Contents

Abstract 1
Disclaimer 2
Table of Contents 3
1   Introduction 6
2   **Building networks** 8
    2.1   **Buildings under consideration** 8
    2.2   **NIST networks and building networks nationwide** 9
       2.2.1   Fire alarm system network 10
       2.2.2   IT network 10
       2.2.3   Heating ventilation and air-conditioning network 13
       2.2.4   Physical access control network 14
       2.2.5   Distributed antenna system in the Advance Measurement Laboratory 15
    2.3   **In-Building Wireless** 16
       2.3.1   Distributed antenna system design 16
       2.3.2   Neutral host approach 17
       2.3.3   Difficulties in implementing dual-use in-building wireless systems 19
    2.4   **Codes and standards** 21
    2.5   **Trends in building network design** 23
3   **Network Requirements** 25
    3.1   **Existing building network application requirements** 25
       3.1.1   IT network 25
       3.1.2   Fire alarm system network 26
       3.1.3   Physical access control network 26
       3.1.4   Heating ventilation and air-conditioning network 27
       3.1.5   Distributed antenna system 27
    3.2   **Building emergency scenarios** 27
       3.2.1   Normal building activity scenario. 28
       3.2.2   Advanced Measurement Laboratory carbon monoxide danger scenario .. 29
       3.2.3   Advanced measurement laboratory large fire scenario 31
       3.2.4   School shooter scenario 34
       3.2.5   Emergency medical service to mall scenario 35
       3.2.6   Scenario summary 35
    3.3   **Public safety requirements of building network** 35
       3.3.1   Incident area network performance requirements 36
       3.3.2   Connecting incident area network to the in-building personal area network 36
    3.4   **Provision of building-source data to public safety users** 37
       3.4.1   Public safety users and building data needs 37

| | 3.4.2 | Scenario requirements analysis .... 38 |
| | 3.4.3 | Additional network requirements .... 42 |

4 Simulation 44

4.1 Incident area network across high-speed building network 44

4.2 P25 interface simulation 47
    4.2.1   P25 communication scenarios 50
    4.2.2   Global statistics 50
    4.2.3   Control end-to-end delay (radio to base) 50
    4.2.4   Voice end-to-end delay 52
    4.2.5   Channel acquisition delay 52
    4.2.6   Message ratio 53
    4.2.7   Analysis 55

4.3 Summary 55

5 Analysis of building roles in emergency response 57

5.1 Extending the incident area network 58
    5.1.1   DAS extension of incident area network 59
    5.1.2   Building network extension of incident area network 61
    5.1.3   Distributed antenna system vs. building network summary 63

5.2 Serving building-source data 64
    5.2.1   Building information services and control system 64
    5.2.2   Getting information to the end user 67

5.3 Recommendations for SoR 69
    5.3.1   Thoughts on the place of the building in the SoR 69
    5.3.2   Overview of submitted SoR change proposals 70

6 Stakeholder interactions 72

6.1 High-rise and large incident communications workshop 72

6.2 NIST SoR Questionnaire 72

6.3 NIST Staff Presentations 72

6.4 Interaction with National Public Safety Telecommunications Council 74

6.5 Interactions with National Electrical Manufacturers Association 74

6.6 Interactions with National Emergency Number Association 75

6.7 Interactions with Alarm Industry Coordinating Council 75

6.8 Discussions with building system vendors and research groups 75
    6.8.1   Honeywell 75
    6.8.2   Siemens 76
    6.8.3   Simplex 76
    6.8.4   Motorola 76
    6.8.5   Johnson Controls 76

6.8.6	University of New Hampshire Interoperability Lab's Project 54	76
6.8.7	United Technologies (UT)	77
6.8.8	NetTalon	77
6.8.9	ISMS	77
6.8.10	Dione Systems	77
6.8.11	Eutech Cybernetics	77
6.8.12	MIJA	78
6.8.13	Bentley	78
6.9	**Stakeholders summary**	78
7	**Summary**	79
8	**References**	81
9	**Appendices**	83

Appendix A: NIST recommended changes to SAFECOM Statement of Requirements 84

    Appendix A.1 SoR v1.1 Section 3.3 Fire-Residential Fire Scenario, revised fire scenario 85

    Appendix A.2 SoR v1.1 Section 4 Operational Requirements of PSWC&I, new Building Data Communications tables 89

    Appendix A.3 SoR v1.1 Section 5 System of Systems, revised including building in network architecture 91

    Appendix A.4 SoR v1.1 Section 7 Public Safety Communications Device Functional Requirements, new "Building Communication System Functional Requirements" subsection 97

    Appendix A.5 SoR v1.1 Section 8.2 Network Functional Requirements, Matrix 96, additional building interface requirements 99

    Appendix A.6 Applications for the Building Information Server to Public Safety Network Interface 100

Appendix B: NIST SoR Questionnaire Feedback 105

Appendix C: Standards for Building Information Exchange with First Responders: presentation to the Alarm Industry Coordinating Council 109

1 Introduction

This report presents work completed under the Building and Fire Research Laboratory (BFRL) Building Networks and Public Safety Communications project sponsored by the Department of Justice via the NIST Office of Law Enforcement Standards (OLES) and in support of SAFECOM efforts to provide a path towards nationwide interoperable public safety radio communications. The main objective of the Building Networks and Public Safety Communications project is to investigate the potential use of the building network infrastructures to facilitate public safety communications. BFRL has conducted basic research related to building network utilization to supplement radio communications.

The project work is divided into three main components: (1) feasibility study of routing public safety communications across a fixed building network infrastructure, (2) providing input to the SAFECOM Statement of Requirements (SoR), and (3) hosting a workshop to address radio communication in high-rise buildings and large/complex incidents. The bulk of the project work is within the first project component. The results of initial research in this first component were used to provide input to the SoR. The workshop was held in June 2006, with a report published separately [Vettori, et. al., 2007] from this report.

The Building Networks and Public Safety Communications project was proposed based on the premise that utilizing building network infrastructures can enable effective communications with emergency responders in large buildings where radio frequency (RF) signals are poor or blocked. Most commercial and institutional buildings have Information Technology (IT) networks, fire alarm system and security networks, and heating ventilation and air-conditioning (HVAC) networks. These networks may provide an effective pathway for transporting mission-critical voice, video and data communications from emergency responders inside the building out to incident command.

In a broader context, the building may have different roles to play in public safety emergency incident response. Based on conclusions of prior work completed under BFRL's earlier OLES-funded project [Jones, W.W., et. al., 2005; Davis, W.D., 2005; Holmberg, D.G., 2006], modern commercial buildings have a wealth of critical information about the conditions within a building that is available through the building automation system that could be used by incident command to help plan effective responses to building incidents.

One may also note that many emergency incidents (at least for the fire service) involve buildings. In addition, large buildings are more complex and are more likely to present RF signal propagation problems. Mitigating this, large commercial buildings likely have sophisticated control networks and may have antenna networks to provide radio coverage within a building. Thus, building information and network infrastructure may play a role in emergency incidents.

Research results show that there is some promise in utilizing building networks to support public safety voice and video communications. Both the information technology (IT) and fire alarm system networks have potential for use in routing public safety communications, but would require some changes. The limitations of these two networks are discussed in Section 2 of this report. Section 2 also gives a review of codes and standards governing buildings that relate to public safety. The use of antenna networks shows more promise for extending public safety communications into buildings. This is discussed in detail in Section 2.3. A summary discussion and tabulated results comparing antenna networks and use of building networks to meet public safety network communication requirements is presented in Section 5.1.

Section 3 presents network requirements based on building emergency scenarios, existing building network application traffic, building codes, and information from public safety practitioners. Section 3 also discusses requirements on the interface between building and public safety network, which includes both the radio interface as well as the connection to the building information systems. Section 4 presents simulation results where the goal was to examine, using simulation tools, the feasibility of routing public safety communications across building networks.

Section 5 presents an analysis of the roles of buildings during emergency response, evaluating the extension of the Incident Area Network (IAN) into the building to support Personal Area Network (PAN) communications as well as the integration with building-source data. This includes the discussion of some work related to an interoperable building data interface for moving building data out to emergency responders (Section 5.2). Section 5 also presents recommendations for the inclusion of buildings into the SAFECOM Statement of Requirements (SoR) document. Section 6 provides an overview of interactions with stakeholders in this work.

The project timeline was structured in such a way as to complete some of the research in time to provide input on the SoR. Our work was submitted, but there was not enough time for the SoR project team to sufficiently evaluate the proposed changes for integrating the building into the SoR networks as proposed. Therefore the BFRL proposed changes were not integrated into the SoR v1.2. This final report represents a complete submission with analysis of the roles that a commercial building may have in public safety response to building emergencies. The project team hopes that the recommendations in this report will be followed and that SAFECOM will consider ways to more fully integrate buildings into the next generation interoperable public safety network.

2 Building networks

This section presents an overview of the current and future state of building networks. What networks exist in large buildings that might be available for carrying public safety communications? How are those networks currently used and for which applications? This section also presents an overview of building codes and standards that may impact the ability to use building networks to carry public safety communications as well as provide an interface to building information systems.

The survey of building networks was begun with an analysis of the networks and network applications in use at NIST. In addition to the NIST study, BFRL staff talked with building experts to extend our knowledge of building networks nationwide and trends within the building controls industry. Literature was searched for additional perspective on present building networks and future trends. Finally, a wide range of building codes and standards were examined.

While the study began with a focus on existing wired building networks, it became clear that the use of in-building wireless (IBW) systems is growing in the large commercial buildings sector. For this reason, some work was done to understand the potential for IBW to serve public safety, and that is presented in Section 2.3. Section 3 focuses on the requirements placed on a building network in order to carry first responder communications.

2.1 Buildings under consideration

What building types merit consideration for use of building networks to move mission-critical voice, video and data? The intuitive response is that buildings with poor RF transmission characteristics are most in need of communications-boosting solutions. This would include larger buildings with more walls. In particular, masonry and steel walls block radio transmissions more than wood stud walls, so larger commercial structures are more likely to have radio transmission problems. But there are other building classes that merit consideration:

- high occupant density: office buildings, hotels, schools;
- high profile: schools, high-rises, government facilities, malls;
- underground spaces: parking garages, basement levels, tunnels; and
- buildings that are not large, but due to design block RF signals. The design could be specifically to block RF, or for other reasons that result in blockage of RF.

One can learn from current local government ordinances what kinds of buildings public safety practitioners are most concerned about. Many cities have adopted regulations mandating that building owners provide public safety radio reception for certain buildings. The recent National Public Safety Telecommunications Council (NPSTC) report [Overby, S., ed., 2007] gives a good overview of the growing number of municipal codes requiring the use of in-building wireless to provide acceptable public safety radio reception as well as the definition of "acceptable". These codes generally cover large buildings over a certain square footage, and explicitly exclude residential structures. The trend is toward continued adoption of similar legislation at the municipal level, with proposals made to consider such legislation in some states. Other publications address this trend [PSWN, 2002; Jacobsmeyer, J.M, 2004]. In addition, coverage of the topic of municipal ordinances is provide at RFsolutions.com.

The argument for supplemental radio coverage solutions provided by the building owner is not clear-cut. Municipalities adopting the type of regulations identified above are primarily smaller cities. Rural municipalities have few problem buildings. Large urban municipalities have many more large structures to deal with and public safety departments generally choose to install a higher density of radio towers to provide a stronger signal that provides better coverage in most structures.

Nonetheless, it remains a truism that some buildings have signal reception problems, and that the ones listed above are the ones most in need of consideration for supplemental systems to provide coverage. It is noteworthy that the trend toward use of IBW systems is focused on the cell phone market and secondarily can benefit public safety. Most modern commercial buildings have sophisticated control systems and networks. Therefore we can examine the potential for using these to meet the need for supplemental radio coverage independently from deciding what kinds of municipalities are most likely to demand this kind of coverage.

2.2 NIST networks and building networks nationwide

In order to understand the potential for using existing building networks for routing voice, video, and other mission critical communications, we proposed studying installed networks, first focused on networks on the NIST campus, and then more broadly looking at building networks nationwide. Studying NIST networks allows us access to information about the various networks, including applications in use, network architecture and traffic profiles, and security.

NIST is a diverse government research institution that includes an Information Technology Laboratory that develops some of the cyber-security standards for U.S. government. It is not surprising, therefore, that NIST has modern networks and takes security very seriously. Information security policies exist for each network on campus and are rigorously developed and followed. In fact, there are four completely separate fiber networks on campus: the life safety fire alarm system network, the information technology local area network (IT LAN), the heating ventilation and air-conditioning network, and the physical access control network. In addition to these there is an independent phone network with a digital PBX and proprietary network protocol that is in process of being phased out and replaced with a Voice over IP (VoIP) solution.

The NIST campus networks are independent of each other, and only the IT LAN connects to the Internet. Consideration was given to using the IT network for physical security, but there was serious concern about opening the access control system to the outside and potential hackers. Likewise, when considering the sensitivity of laboratory experiments and required environmental conditions, the decision was made to keep HVAC network traffic separate from the IT LAN. Security measures went as far as keeping HVAC network devices in separate closets from IT network devices.

In the course of this study, we spoke to the network administrators of each of the networks, examined policy documents, architecture diagrams, toured key network equipment rooms, and discussed application traffic, protocols, and security. This section summarizes this information for the NIST networks and then applies these results nationwide.

2.2.1 Fire alarm system network

The fire alarm system network has a fiber optic backbone loop connecting to the fire panel in each building on campus, with a communications center at the fire station on campus. Within each building, the fire alarm system architecture uses digitally addressed fire sensors and annunciators connected on a low-speed network. The network operates normally with 30 V signaling which increases to 36 V signaling in alarm. The wiring is twisted pair with more than double the capacitance of standard ANSI/TIA/EIA-568-A Category 5 cabling, and T-tapped style connections to devices which result in uncontrolled impedance. The resulting low speed on the network (3300 bps) makes it unsuitable for high-bandwidth video applications and even lower-bandwidth voice communications.

However, the fire alarm system network has many advantages that suit first responder communications if a next-generation fire alarm system network were built that allowed high-bandwidth data transfer. These network advantages arise from the requirements placed on the network by the fire code. The National Fire Protection Association Standard 72 (NFPA 72, *National Fire Alarm Code*) requires: guaranteed supervisory and alarm signal delivery within a set number of seconds, signal path integrity monitoring, redundant signal paths, backup power, physical protection, and tolerance of humidity, temperature, and voltage extremes. The network is under the control of the local fire authority and is also reliable and thus already trusted by public safety practitioners. Finally, the fire alarm system network extends to every space in the building, occupied spaces as well as service areas (attic, mechanical, parking garages) and thus is readily accessible for connecting wireless access points (AP) in all these areas. However, the need for code changes before implementing such a system adds to the difficulty of transitioning the fire alarm system network for new uses.

2.2.2 IT network

The NIST IT network is separate from other networks described in this section, and is the only network with connection to the Internet. It is a modern, high-bandwidth general purpose TCP/IP network that will handle all the voice and video traffic that could be generated during a large building emergency incident, if those emergency communications were to be transported over some part of the IT network.

The current NIST IT network consists of redundant 2 Gigabit per second (Gbps) fiber backbones that run in a loop connecting the buildings on campus. At each building there is a main router with 1 Gbps connections to a number of switches in data closets throughout the building. Each switch has 10/100 Mbps coaxial cable running to personal computers. Available bandwidth is currently much greater than what is needed. The NIST Office of the Chief Information Officer makes network traffic data statistics available on its internal website, where it shows peak network usage is on the order of 5 % of capacity, whereas average bandwidth in use on the backbone is less than 1 % of capacity.

NIST currently does not provide wireless network services. The NIST policy has been to approach the issue carefully and avoid opening the internal network to outsiders who could gain access at the NIST perimeter. However, NIST is planning to provide wireless service, waiting on a more developed system for user authorization to allow association of computers with IP addresses using dynamic configuration.

In general, NIST is building up network capabilities in preparation for wireless, IPv6 transition, and phone system integration using VoIP. The federal government was given a June 2008 Office of Management and Budget mandate to have networks IPv6 "capable". Equipment is being upgraded to allow the network to run IPv6 applications alongside current IPv4 applications. For the transition to VoIP, significant changes must be put in place. These include priority routing for phone system packet quality of service (QoS), power backup for the phone system, and in-line power (using power over Ethernet) to phones. The actual bandwidth used by the phone system is negligible, since voice traffic only uses 64 kbps (this is a maximum value for the H.323 which is the leading VoIP protocol [Desourdis, R.I, 2002, p.184]) of the available 100 Mbps line to the data closet, and cumulative traffic on the backbone will consume a small percentage of available capacity.

The IT network can clearly handle the bandwidth requirements of emergency responder voice, video and data communications. The IT network is available in all occupied spaces for attaching wireless APs, but would need extensions to reach service spaces (attic, mechanical, parking garages) required for emergency response. However, unlike the fire alarm system network, the IT network has no code requirements concerning security, change management, redundant cables, backup power, and fire resistance, and thus is likely to be less reliable.

The *National Fire Alarm Code* (NFPA 72) has requirements for public systems that are used for transmitting fire alarm signals. The requirements include a written contract governing control of the network, with the local public safety authority having anytime access to the network. All maintenance to and changes on the network are logged, and testing of components is required monthly. If emergency responders are to rely on the building IT network for routing mission critical voice and other traffic, then standards would need to be established to govern network management, performance requirements, and additional testing and reporting.

When considering IT networks nationwide, there are some differences and trends to consider. In general, large commercial facilities will have more resources to support the IT network. This includes dedicated personnel, with resulting sophistication in policy, and generally better security and reliability. Small commercial facilities may have a simple IT network supported by a single person, or no IT network (similar to residential). Large corporate facilities, hotels, hospitals likely have IT networks that could support public safety communications traffic. However, many facilities of interest to public safety lean toward the small commercial facility: schools, malls, and other special facilities. Leased office space likely falls into this category as well, since each tenant maintains its own IT network. The building owner would have to control and manage the building IT network in order for the IT network to route public safety communications.

When there is a sophisticated IT network with careful management, there will also be management concerns related to external control and access to the network, which would be required for the public safety application. Every device attached to the network from outside the firewall is essentially a door into the IT network, and the network administrators must be concerned about additional security issues. On the public safety side, local public safety authorities will want access to test and monitor the public safety communications system to ensure that it is operational.

The IT network is by nature general purpose. There are many situations that may lead to network failure. Some examples include: applications on the network that undergo some upgrade leading to unexpected network behavior such as flooding with unwanted messages; Internet-source security issues such as a denial of service attack; or a physical network issue such as a technician disconnecting a cable or connecting it improperly. The fire alarm system network has code controls to minimize any of these problems, but the reliability of the IT network is much more sensitive to the quality of the IT department, and subject to some degree to uncontrollable factors. Public safety departments may not be able to accept this, and IT network administrators may likely not want to give enough control to public safety jurisdiction.

There is a trend in general toward integrating networks in large commercial buildings. Just as NIST is working toward integrating telephone voice on to the IT network, many new buildings have consolidated the networks such that the IT network also supports HVAC and physical access control. Fire alarm system networks will likely be the last to be integrated due to the stringent demands for reliability and UL certification, but even this may occur. One good example of the movement in this direction is Cisco Systems' Connected Real Estate. Cisco sees the benefit not only of sharing physical infrastructure, but also information between systems to gain enterprise-wide awareness and efficiencies [http://www.cisco.com/web/strategy/trec/index.html]. The good news is that the technologies and protocols being used to segregate and protect individual system traffic streams may allow for guaranteed priority and quality of service for public safety communications routed across the network.

2.2.3 Heating ventilation and air-conditioning network

The HVAC network on the NIST campus is isolated from other networks in order to better protect the mechanical equipment that controls the sensitive laboratory environments. Like the other networks, the HVAC network has a fiber backbone that runs in a loop (actually several physically separate loops serving different sets of buildings) with building main routers that link to building equipment controllers in the mechanical spaces.

The different HVAC network loops are linked together at the Plant building where plant division personnel monitor and maintain the network. Network administrators tightly control user authorization using role based access control to limit the authority of different maintenance personnel to change device configurations that in doing so could seriously affect the operation of environmental controls.

The HVAC network has high-speed connections between building controllers, with lower-speed networks extending to lower-level field controllers and end devices. While the high-speed network is limited to mechanical room space (attic), the lower speed network extends in some newer buildings down to variable air volume controllers throughout the building. In the case of the older buildings, the mechanical system uses mechanical controls at the room air vents so that the HVAC network is confined to the attic mechanical space.

A study was performed on the NIST HVAC network in order to understand the kinds and amount of HVAC network traffic in a building representative of some existing commercial buildings. Two different software tools were used to examine the different kinds of traffic on the network, allowing for packet filtering and examination of BACnet and other message types. These tools have interfaces that allows sorting and viewing of packet headers, as well as the packet raw data in hex format, and also statistics of packet type and packet per second on the wire.

These two tools were installed on a computer connected to a hub in the attic of the Building Research Laboratory. The hub connects the computer and four main controllers in the attic to a switch (via cat 5 cable and using 100 Mbps capability) linking the building controllers to the HVAC network fiber backbone. Of the four controllers in the building, three are Andover controllers that operate using BACnet/IP (B/IP), and one is an Alerton controller (Alerton BTI) which uses BACnet over Ethernet (B/Eth). This Alerton BTI also functions as a campus wide B/IP to B/Eth router. Therefore, any communication between Andover and Alerton controllers anywhere on campus gets routed through the Alerton BTI router.

With this architecture, the traffic seen at the research computer connected to the hub in the attic includes any traffic to or from BACnet devices in the Building Research Laboratory (the four main controllers as well as any devices on subnets behind them) plus any broadcast traffic, plus any traffic routed from IP to Ethernet or vice versa.

HVAC network traffic was collected at various times for different time durations in order to better understand the HVAC network traffic. The traffic volume was found to be very low (less that 20 kbps average) compared to the capacity of the network (100 Mbps in the individual buildings, 1 Gbps between buildings on the backbone). The traffic itself was composed largely of BACnet messages, with Change_of_Value notifications the most common, followed by Read Property messages. At the time of the study, only 26 devices were directly connected to the network, consisting primarily of facility air handling unit controllers but also several workstations used for supervisory control. However, a larger number of BACnet sub-controllers connect to subnets behind many of these main controllers, and some of the messages on the backbone originated or terminated at devices on these subnets. The final results demonstrate that there is plenty of bandwidth available for non-HVAC purposes, such as routing of emergency responder voice and video traffic.

In older NIST buildings the control network does not extend beyond the attic space. In newer buildings, lower-speed sub-networks extend from the attic controllers to sub-controllers which in turn connect to individual control points. These lower speed sub-networks route controller communications over twisted pair wiring with a datalink protocol having baud rates typically in the 100 Kbps range and thus could support voice and some data transmissions, but would not support streaming video applications.
In addition to bandwidth limitations, the HVAC network does not extend to all spaces in the building, and faces the same management concerns as with the IT network. However, because the HVAC network has fewer applications and users, it may be more reliable than the general purpose IT network.

Outside NIST, as mentioned above, it is increasingly common to have higher level HVAC controllers sit on the IT network. The higher level controllers then serve as routers to lower speed controls networks that serve different zones of the building. Potentially the higher speed network can be extended to the end control points. One model that seems to be gaining ground is that of a single high-speed IP network that serves all building network communications: IT, mechanical control, security (physical access control), telecommunications, and perhaps others. As this trend grows, any network communications sharing the converged network are then subject to the limitations of that network.

2.2.4 Physical access control network
As with other NIST networks, the physical access control network is isolated from other networks. A fiber loop connects the different buildings, with the main network control located in the NIST security office. NIST security officers can monitor the activity of all building and internal zone access points with alerts issued by card readers, and camera views at sensitive entry doors.

The physical security system is a Lenel system, and each NIST campus building has one Lenel controller linked via a switch to the fiber backbone. Each controller has EIA-485 wiring (OSI Model physical layer electrical specification of a twisted pair multipoint serial connection that allows for data transfer over long distances) to all the readers and door hardware with a total of 230 readers on campus, most of them basic card readers, but some with other or additional authentication such as pin pad. Data speed on the EIA-485 network is limited to 38.4 Kbps. Communications from the building controller to the campus security center use TCP/IP connections across the fiber network. All system communications use Advanced Encryption Standard (AES) encryption [Lenel, 2007]

Card readers scan cards and pass card identifiers to the door controller which has a local copy of the NIST-wide user authorization data. The door controller passes back an allow/disallow decision. Traffic on the backbone consists of alarms to the security center, and user data auto-updates from security center to each controller. User data includes which users are valid on which dates and times of day.

While the NIST access control network physical wiring could theoretically support voice traffic, the protocol in use by the Lenel system is proprietary. The Lenel controllers would have to be redesigned to handle and route first responder communications traffic. In addition, the current access control network is physically only available in a limited number of locations in any building—wherever access control is required. This generally is limited to external doors or doors to sensitive areas, typically only a few locations in the building.

However, the trend nationwide is toward moving physical access control on to general purpose IT networks. This is made possible by the use of standard encryption algorithms available on today's modern IP networks. In addition, virtual LAN and other technologies allow for greater control and separation of traffic streams on a converged IP network.

2.2.5 Distributed antenna system in the Advance Measurement Laboratory

The NIST Advanced Measurement Laboratory (AML) has two buildings entirely underground. Radio reception in these buildings is poor to none. NIST has implemented a solution that gives NIST fire and police radio reception during emergencies (normally the system is off since the desire is to prevent radio signal propagation in the buildings).

The solution is tailored to the radio frequencies used for Montgomery County, MD, in the 800 MHz public safety band. A distributed antenna system in selected lower level hallways and stairwells provides for radio coverage when the system is powered up during emergencies or testing. The solution is implemented using a directional antenna above ground that is directed at the local public safety land mobile radio base station. This antenna is connected via coaxial cable to a bi-directional amplifier (BDA) within the building complex. The amplifier filters out unwanted frequencies (outside the narrow 800 MHz bands). Fiber-optic cables carry signals to and from remote building hubs, which then have coaxial cable running to antennas in areas with required reception.

The use of distributed antenna systems as a solution for in-building radio reception is discussed in Section 2.3.

2.3 In-Building Wireless

This section addresses the issues related to use of in-building distributed antenna systems (DAS) and repeater systems to connect first responders inside a building to the Incident Area Network on the exterior.

The fundamental purpose of in-building wireless systems is to extend strong radio signals inside buildings or public venues such as airports, stadiums, parking garages, subway stations, and tunnels. Cellular carriers use networks of outdoor base stations to distribute their signals to outdoor users, but building materials such as concrete and steel attenuate these signals. As a result, cell phones often work indoors only when near the exterior walls, but they do not work well or at all as users move into interior offices, stairwells, or underground parking garages. The same can be said for public safety radios. In many large buildings radio reception in parts of the building may be non-existent, particularly interior and underground spaces.

One method to overcome these difficulties is with the installation of a distributed antenna system. A DAS brings the radio signal into the building via antennas distributed throughout the interior space. A DAS may use long runs of "leaky coax" cable extending down corridors in a building to propagate and collect radio signals in desired areas of a building, or may use discrete antennas (such as for IEEE 802.11 WiFi) attached to a wired network. A DAS can provide good indoor coverage for wireless communication systems without transmitting at very high powers. Currently the DAS concept is primarily used in conjunction with in-building wireless systems designed to provide service to cell phone users and for providing wireless Internet access.

2.3.1 Distributed antenna system design

Distributed antenna systems have been widely implemented in state-of-the-art cellular communication systems to cover dead spots. In this case, increasing the number of cells has the benefits of better coverage (fewer dead spots) at lower power and increased capacity. The idea works because less power is wasted in overcoming penetration and shadowing losses, and because a line-of-sight channel is present more frequently, leading to reduced fade depths and reduced delay spread. A building is effectively a dead spot in the cellular system. An in-building wireless system uses a DAS to establish either a single cell (the antenna tree), or multiple cells (the WiFi access point approach) that effectively extends coverage to all desired areas of a building. In-building wireless systems fall into two main categories: passive and active.

2.3.1.1 Passive systems

Passive systems use rigid coaxial cable to distribute the wireless signal from a repeater or base station to a set of distributed antennas. These systems may use coaxial couplers or splitters to achieve the proper geographic distribution of cabling, but these systems are passive because the distributed antenna system itself uses no electronic components. The antennas can be an extended portion of leaky coax, or the typical omni-directional antennas at discrete locations. Passive systems have the disadvantage of limited geographical size, but the advantage of a wider frequency range since there is no reliance on amplifiers that are frequency band specific. Antennas in different parts of the building are joined in a tree architecture with a radio receiver at the base of the tree. The radio receiver can then route the signal over a wired digital connection to a base station or repeater, or rebroadcast the signal on an externally mounted antenna.

2.3.1.2 Active systems

Active in-building systems use electronics that amplify the signal for distribution. In many cases, an active system uses fiber running up a building riser to link a main hub with expansion hubs, and then uses CAT-5 cabling to connect each expansion hub to a remote access unit and connected antenna(s). An active system might also be used in a campus environment to connect antennas in several buildings back to a central location. An active system generally passes certain desired frequencies and filters other undesired frequencies.

There are systems that are a hybrid of these two, featuring active electronics in the form of a head-end unit connected via fiber home runs to remote units, but using passive coaxial cabling to link these remote units to antennas. But, in essence, all antennas are passive. The question is where the radio receiver/source is and whether there are intervening electronics.

2.3.2 Neutral host approach

The trend in the building industry is moving toward providing in-building wireless access. Evidence of this can be seen in the formation in 2006 of the In-Building Wireless Alliance, who's members include companies in the fields of real estate, building controls and wireless communications. Their stated goal is to make the business case for in-building wireless systems in order to build the market for these systems. From a non-public safety perspective, the IBWA has done market research and concluded that in-building wireless systems will give a high return on investment to the building owner due to the willingness of tenants to pay for such features as ubiquitous cell and WiFi service.

To best serve the different radio applications (e.g., different cell phone carriers, WiFi) building owners are commonly implementing "neutral host" in-building wireless systems. In these systems, the DAS and electronics are designed to provide radio coverage for a wide frequency band that covers typically the 800 MHz and 1900 MHz cellular bands. The DAS is then made available to outside cellular carriers to provide service to their networks by installing a base station in the building.

One example of this is the DAS in place at Hartsfield-Jackson Atlanta International Airport, which recently placed into service two wireless networks that cover all concourses and gates [Cox, J., 2006]. The system uses fiber, shielded coaxial cable and hardware to distribute cellular signals from carrier base stations to ceiling-mounted antennas throughout the airport. This distributed antenna system infrastructure can also be used by city, state and federal public safety staff and emergency responders with radios using the dedicated 851 MHz to 869 MHz band.

Another example of a neutral host system is the "wireless utility" marketed by Johnson Controls, Inc. using passive distributed antenna systems. The "wireless utility" building product is essentially a neutral host DAS that spans the range from 400 MHz to 2500 MHz and 4.9 GHz up to 5.8 GHz and thus covers frequency ranges from cellular up to IEEE 802.11a WiFi, and including common public safety bands.

2.3.2.1 Existing Public Safety in-building radio coverage

In contrast to the neutral host concept, which is sold to building owners as a profit generating utility, public safety in-building radio coverage is not a profit center and is provided as required by municipal ordinance or in response to special request by public safety officials. As noted above, a neutral host cellular in-building wireless system can be used to provide service to public safety officials, if the public safety radios operate on a frequency band that is supported by the neutral host system. Many of the neutral host systems support the 800 MHz public safety bands (commonly used for many conventional and trunked systems) because they are adjacent to cell phone spectrum. Some neutral host systems even support a wider band extending to lower frequencies, but there are in fact many public safety bands in use: Very High Frequency (VHF) low bands at 20 MHz to 50 MHz, VHF high bands at 138 MHz to 172 MHz, Ultra-High Frequency (UHF) low bands in the 400 MHz range, the new upper 700 MHz range bands, and the 800 MHz bands [Desourdis, 2002, Tables 2.1, 2.2]. In addition, the areas of a building for which public safety access is required includes spaces such as utility areas that a neutral host system would not cover unless specifically designed for public safety needs. This issue will be discussed more fully later.

Another fundamental difference exists between the cellular architecture and the common land mobile radio (LMR) system architecture, and that is the need for wide area talk groups. Whereas cellular architecture can increase capacity by decreasing cell size, public safety radio use requires sharing of frequencies by talk groups dispersed over some geographical area. This results in an optimal architecture with few cells and resulting few base stations broadcasting at higher power. Thus, it is not especially helpful to have an additional base station in place within a building. Instead it is common to have a repeater arrangement where the signals within a building are moved to the outside and rebroadcast with a directional antenna directed toward the closest base station. For building areas where a portable radio can receive the base station signal but lacks power to transmit back to base, a voting/satellite receiver can be used to pick up the portable signal and repeat the amplified signal to base. For building spaces that have no reception, it is common to use a gap-filler repeater, or bi-directional amplifier (BDA, also known as signal booster) that has internal and external antennas. Because of physical separation inside to outside the signal received on the external antenna can be rebroadcast on the interior on the same frequency without feedback interference.

The NIST Advanced Measurement Laboratory is a good example of a facility needing gap-filler support. The entire structures of two of the laboratory buildings are underground with no radio signals. In fact, the buildings were purposely built underground in order to block radio noise. However, when an emergency occurs, a DAS connected to a BDA can be powered on to allow first responders to have radio communications on the local 800 MHz trunked radio network.

2.3.2.2 Municipalities requiring in-building public safety radio coverage

It is becoming common now for municipalities to use local ordinances to compel large building owners to provide a minimum level of radio reception throughout a minimum percentage of a facility [www.RFsolutions.com]. These ordinances are moving toward a standard format as cities have gained experience and share information amongst themselves. The National Public Safety Telecommunications Council (NPSTC) report [Overby, S., ed., 2007] presents a review of municipalities with such ordinances and a summary of the ordinances' provisions. An analysis of the content of various municipal ordinances can also be found on the RFsolutions.com website.

If a municipality requires building owners to install in-building wireless systems to support public safety needs, and building owners are simultaneously considering installing in-building wireless systems to support tenant communication needs, then what efforts can be made to join these two trends? The answer is that the public safety community should be working with groups like the In-Building Wireless Alliance to find ways to enable public safety communications to share the in-building wireless infrastructure, piggy-backing on the larger society trend and making the business case for the building owner stronger, and allowing the building owner to offer tenants not only cell phone and WiFi but also increased safety. The NPSTC report provides more guidance on best practices for implementing in-building wireless systems to serve public safety needs.

2.3.3 Difficulties in implementing dual-use in-building wireless systems

There are many challenges to using a neutral host in-building wireless system to serve public safety radio needs. This section reviews these challenges.

2.3.3.1 Public Safety in-building requirements

The public safety requirements for in-building radio coverage include:
1. provision of a specified minimum signal strength to specified regions of a building with a certain percentage of coverage and with no change over time [Overby, S., ed., 2007];
2. coverage of specified regions of a building such as parking, stairwells, and utility areas of building;
3. handling frequency ranges of existing public safety communications;
4. handling the power levels of mobile units (i.e., in parking garages) which are higher than typical cell phone handsets [Overby, S., ed., 2007];
5. allowing for connecting building interior to exterior and different repeater arrangements as needed;
6. allowing public safety officials to control implementation enough to guarantee reliability and functionality to meet public safety needs (see Section 2.3.3.5).

Some of the challenges of using a neutral host system for meeting these public safety radio coverage requirements are addressed below.

2.3.3.2 Coverage

A review of municipality ordinances [PSWN, 2002] showed that some required a minimum signal level in buildings of –107 dBm, while about as many required a stronger signal of –95 dB. Also, the percentage of building floor space required to have this minimum coverage varied from 85 % up to 95%. Jack Daniel [Daniel, J., 2005], makes the following statement regarding recommended signal coverage, "many system engineers recommend –95 dBm or -100 dBm signal levels, 95 % of the area and 95 % reliability as a proven, achievable and measurable balance of function versus cost."

Neutral host solutions may not provide this coverage. They almost certainly would not be designed to provide coverage in utility areas, and may not provide coverage in stairwells and underground parking areas. Therefore, any building owner required (or desiring) to provide public safety coverage with a shared (neutral host) in-building wireless installation will need to begin with the public safety coverage requirements and meet the union of the non-public safety and public safety demands.

2.3.3.3 Frequency bands, interference, power levels

A building owner must provide an in-building wireless system that meets the requirements of the local, state, and federal public safety radio frequency bands in use in the local area, or at least those bands that public safety officials judge as most important. A review of available systems and spectrum in use by public safety applications indicates that most public safety communications exist in the 400 MHz and 800 MHz bands while some legacy systems still operate in the 150 MHz and lower bands. It is not clear that any neutral host system vendors offer service at frequencies below 400 MHz. Additionally, if an active system is in use, then provisions must be made to amplify all required public safety frequency bands.

A problem may arise for signals in bands that are closely spaced and border other bands in use in the building for cellular service. In this case special provision must be made for band-pass filters that insure neighboring bands are sufficiently filtered to prevent interference on the public safety channel. As for power level issues, any in-building wireless system must be able to handle the higher signal power levels of public safety portable handsets and even higher levels of mobile units (in parking garages) without distortion of the signal.

2.3.3.4 Use during emergencies

Beyond maintenance concerns, there are also design concerns that provide for: redundancy, power backup, and survivability [Overby, S., ed., 2007]. Codes and ordinances may be put in place with authority references to the fire code, and the fire code has provisions for redundancy, power backup, and survivability. It is important to note that a typical for-profit neutral host solution that does not consider public safety applications would not generally provide the redundancy, power backup, and survivability required.

2.3.3.5 Ownership and maintenance

Public safety officials rely on systems in buildings, such as the fire alarm system, that they do not have authority over nor sole use of. Other systems used by public safety officials include the paging system and smoke control system. Can emergency responders rely on an in-building wireless system shared with cell phone and other building occupant used services? Who has authority to modify the system to add a new service, relocate an antenna, or replace/ upgrade electronics? Jack Daniel [Daniel, J., 2005] makes the point that first responders must have (a) unlimited access to the system 24/7 to disable interfering equipment or implement changes to enhance coverage during an emergency, and (b) systems not altered without consent.

There must be policy controls to ensure that public safety needs are not disserved. This can be accomplished via different means:
- Written policy approved by building owner and Authority Having Jurisdiction stating who can make changes, and testing procedures to ensure public safety requirements are still met after changes are completed; and,
- Standard municipal requirements codifying the same. This would be part of the code that mandates in-building coverage. The white paper by Jack Daniel [Daniel, J., 2005] provides guidance on common and necessary elements of a municipal code governing in-building wireless systems for public safety.

It is significant that the Public Safety Wireless Network (PSWN) in-building ordinances report notes the ability of municipal ordinances "to effect the development of in-building wireless systems that mitigate or resolve the problem of public safety in-building wireless access." [PSWN, 2002]. The NPSTC report [Overby, S., ed., 2007] gives an update on national model code initiatives at state and national levels as the various codes of many local districts are examined for broader application. A report to the Virginia House of Representatives [VA House, 2003] states in summary that, "The basic principles governing public safety radio systems are stable enough, however, that the installation of emergency communications equipment in certain buildings to provide effective and reliable communications for emergency response personnel need not be postponed." Jack Daniel's review of municipal ordinances in 2005 makes a stronger statement.

2.4 Codes and standards

The purpose of this section is to review standards that relate to moving mission critical voice and video across building networks. Building codes govern the fire alarm system network. Some common standards exist also for communications on the facilities and security networks.

The International Building Code (IBC) has been adopted by most states and provides the minimum requirements for fire safety and other hazards for the built environment. It also provides for the safety of fire fighters and emergency responders during emergency operations. Other codes such as the International Fire Code and the International Code Council (ICC) Electrical Code are considered part of this code as well as referenced standards and codes from the National Fire Protection Association (NFPA), such as

NFPA 13 (sprinklers), NFPA 72 (detection) and NFPA 101 (life safety). Some states will use NFPA 1 as their fire code and include NFPA 13, NFPA 72 and NFPA 101 but still use IBC as their building code. The application of these codes depends on the building code officials and fire marshals of a particular city or state.

An emergency voice/alarm communication system is required by the IBC for high-rise buildings defined as 23 m (75 ft) above the lowest level of fire vehicle access, for atriums connecting more than two stories, and for underground buildings where the lowest level of the structure is more than 18 m (60 ft) below the lowest level of exit discharge. Speakers for this communication system will be provided at a minimum for elevator groups, exit stairways, each floor, and areas of refuge. The system will have the capability of broadcasting live voice messages.

For high-rise buildings, a fire department communication system ('fire phones") is also required. This two-way communication system is for fire department use and connects the fire command center with elevators, elevator lobbies, emergency and standby power rooms, fire pump rooms, areas of refuge, and inside enclosed exit stairways. The fire department communication device shall be located at each floor level within the enclosed stairway. Potentially these already-required communications systems could be designed to also carry voice and video communications.

NFPA 72 (2007) contains a number of important standards that in-building communication equipment must meet in order to interact with building fire alarm systems or be considered as reliable as a fire alarm system. The following sections/chapters are relevant.:
1. Chapter 4, Fundamentals of Fire Alarm Systems, provides guidance as to the credentialing and qualifications of personnel who design and install the equipment. Also included in the chapter are cabling requirements and equipment power requirements:
    - *Power Supplies* – 4.4.1 for protected premises fire alarm systems and supervising station facilities. Primary power provided by commercial light and power or an engine driven generator. Backup power must have sufficient capacity to operate the system for 24 hours and then be capable of operating the system under full load for either 5 or 15 minutes dependent on the type of system.
    - *Performance and Limitations* - 4.4.4 requires that equipment be designed such that it is capable of performing at ambient temperatures of 0 °C to 49 °C and at a relative humidity of 85 % at 30 °C. This section also requires that equipment be capable of operating at 85 % and 110 % of the nameplate primary and secondary input voltages, and also requires that all wiring, cabling, and equipment installation be in accordance with NFPA 70, National Electrical Code.
2. *Requirements for Smoke and Heat Detectors* 5.5 requires coverage everywhere although there are notable exceptions. This implies that the fire alarm system network will be present in all building areas of interest to public safety.

3. Interconnection with non-fire alarm systems is permitted in 6.8.4.1 and described in detail in Annex A6.8.4.1 and A6.8.4.2.
4. Chapter 8, *Supervising Station Fire Alarm Systems*, provides some useful guidance for the requirements of a system sending alarms to a remote site. Some of these requirements include the need for alternate transmission paths, a 90% probability of successfully completing each transmission sequence, need for periodic test signals, and maximum allowable transmission times.
5. Table 10.4.2.2 item 19, *Supervising Station Fire Alarm Systems*, item 23, *Fire Safety Functions*, and item 26, *Low-Power Radio (Wireless Systems)* provide the guidance for inspection, testing, and maintenance.
6. Annex F, NEMA SB 30 *Fire Service Annunciator and Interface* contains an extensive list of relevant standards and codes and references to wireless applications and remote access for the fire service. The annex is not part of the requirements for NFPA 72 but is included for informational purposes. It was developed by NEMA as a standard to guide manufacturers in the development of uniform equipment for the fire service.

Codes and standards that are not included in the building code but have been developed in the consensus based arena by industry associations, testing laboratories, and standards groups are relevant. NFPA 731 Standard for the Installation of Electronic Premises Security Systems covers installation, testing, inspection and maintenance of electronic security equipment. This standard also covers the connection to central station alarm companies and verification of alarms. The Security Industry Association ANSI/SIA DC-09-2007 Digital Communication Standard – Internet Protocol Event Reporting details the protocol and related details to report security events using Internet protocol (IP) to carry the event content.

Building facility (mechanical) networks have BACnet (ASHRAE 135-2004) as the open standard communications protocol that allows a wide variety of devices to communicate with each other over IP, Ethernet and other networks. Other proprietary communication protocols exist and can be connected to BACnet networks via gateways. The BACnet standard is being expanded to allow fire and security systems to operate with the BACnet protocol, and BACnet can be used on an IP network. It seems likely that the use of BACnet will continue to grow and be of greater value on a converged IP network.

2.5 Trends in building network design

There are a number of trends in building network design which may impact the potential for building networks to serve public safety. These trends have come up in discussions with building controls experts, as well as seen recurring in building networking publications.

Network convergence
As noted earlier, there is a trend toward converged networks. Building security networks are already being commonly implemented on the IT network. Higher level HVAC controllers are often sharing the IT backbone. At some point in the future, the fire alarm

system network may also move toward sharing an IT network if code requirements can be satisfied. This move to a single high-speed facility network will provide a better platform for serving public safety needs for a high bandwidth network that extends throughout a facility.

Wireless networks and protocols
Wireless networking is growing in use in many areas, including building networks, although still in its infancy there. IEEE 802.11 WiFi is already firmly established for computer networking, while IEEE 802.16 WiMAX, and IEEE 802.15.4 Zigbee seem to be growing in use. ZigBee is targeted at applications that require a low data rate and low power consumption, and with secure networking. Building network vendors are looking at using WiFi and Zigbee in HVAC and building security networks. From a building security network perspective, some issues that still need to be resolved include end-to-end security, power consumption, and redundancy to prevent jamming of signals taking down the network.

Distributed Antenna Systems
There is a trend toward use of distributed antenna systems for provision of radio access inside buildings, as discussed more fully in Section 2.3. The use of DAS has many benefits beyond the typical application of providing cell phone service. A DAS can extend radio signal range into a building for a certain design range of frequencies. "Leaky coax" passive antenna systems also effectively increase the range of wireless devices and thus support mesh networks (the antenna carries the signal of one device around obstructions and thus extends signal range).

More robust networks
Homeland security and disaster preparedness is driving a trend toward more robust networks, at least for public safety and other critical functions. More robust networks require better IT security, better power backup, more physical security in some facilities, and potentially more sensors for monitoring the network and surrounding environment.

3 Network Requirements

This section of the report addresses the demands on the building network during a building emergency and includes both the demands placed on the network by existing network applications as well as the demands placed on the network by emergency responder voice, video and data. This is addressed first by looking at existing building network applications and secondly by analyzing building incident scenarios and demands placed on building networks by communication needs of responders during these incident scenarios. Finally, public safety communication scenario demands on the network and the requirements on interfaces between the public safety network and the building network are examined to determine requirements for a building-source data interface.

3.1 Existing building network application requirements

Each of the networks discussed earlier in Section 2.2 have applications that place demands on the building networks. Here we review those demands, including bandwidth from application traffic, both regular traffic as well as during a building incident.

3.1.1 IT network

Normal applications running on the corporate IT network include all network traffic, some of which originates with a personal computer (PC) application. This includes application traffic that is user initiated as well as regularly scheduled application traffic. In addition to PC traffic, there will be traffic that originates at other network device nodes.

Personal computer applications run by users include email, Internet browser, word processing, spreadsheet and database applications, webpage development, and many more specific tools such as drafting, video editing, scientific data processing, etc. These may or may not produce any significant network traffic. Email and web applications clearly produce network traffic, with some web applications (such as streaming video or file transfers) producing larger amounts. Many PC applications may produce network traffic because the software is connected to a central server-based database. This is the case for many NIST secretarial applications such as travel management, time sheet recording, as well as other employee applications such as email and corporate calendaring software.

The applications mentioned above are all opened by users and network traffic is initiated by the actions of users at their PC. Therefore, in an emergency, when people evacuate, these applications would stop generating new traffic on the network.

Some other network traffic is controlled by applications that run autonomously on some preset schedule. This includes backup services where a daily data transfer is made from employee PC to central server, and operating system or other software upgrades/patches. There are also network applications that are independent of the PC: building automation system and access control system communications that share the IT network, VoIP communications, etc.

All non-user initiated communications will continue on the network during an emergency event, unless they are specifically programmed to shut down. The VoIP system will likely see a drop in traffic if the phones are desk top handsets as users evacuate. Network backup of PC data to a central server and other large data transfers are applications that consume significant amounts of bandwidth. In the NIST case, data transfers were higher at night and produced network loads of only several percent of capacity. The NIST network as a whole generally operates at several percent of capacity.

3.1.2 Fire alarm system network
Current fire alarm system network traffic is very low bandwidth. During normal operation, the only traffic on the network are steady heartbeat messages to confirm that all network devices are up and operational. Any problems are reported to a human user. During an emergency the system bandwidth is still low with the addition of alarm messages from specific devices (smoke alarm, heat sensor alarm, pull station alarm) to the fire panel indicating an alarm status. The fire panel will respond by turning on the annunciators (alarm bells) in the building, and also by sending a message to an alarm company indicating that there has been a fire alarm.

The fire code places requirements on the fire alarm system network as outlined in Section 2.4. Some of these requirements specifically address the interconnections allowed with other networks. Any plan to route emergency communications traffic over the fire alarm system network will need to comply with these provisions of the fire code or else changes must be made to the fire code.

3.1.3 Physical access control network
Physical access control network communications are typically low bandwidth. Generally, the system is configured such that badge swipes or other credential reader actions only generate traffic between the reader and the door controller. The only traffic on the building network are regular user database transfers from a central database to each local controller so that each local controller has a copy of the master database which identifies which users have access to which zones using which credential at what times, etc. The only potentially large application is that of video traffic between remote cameras and a central station where it is viewed. Some current systems store video signals centrally, which creates much more network traffic. More modern systems store the digital signal near the camera and then if requested will send the video signal over the network to a requesting user.

During an emergency, it is likely that video cameras will be used, but traffic may not be more than at any other time since cameras are generally monitored at all times from a central security station. Any camera in the area of the incident will likely be very valuable for incident assessment, and the network should be capable of supporting live video from each camera. However, for many scenarios (such as fire), a low frame rate signal may be acceptable. In addition, for an intruder scenario it may be acceptable to have a low-resolution high frame rate video signal, although high-resolution frame capture could be important for intruder identification or other purposes. Storage of all high-resolution video signals should be provided, preferably at a location removed from the incident (i.e., away from a fire).

3.1.4 Heating ventilation and air-conditioning network

The building automation system continues operation during an incident, unless there is some human intervention. HVAC may be commanded to a different mode such as for smoke control, or may be shut down. Lights, elevators, access control, and any other systems with controllers on the building automation system network, are not likely to experience significant changes in use during an emergency except from normal mode to some egress mode. In general, there will be no high-bandwidth traffic due to building automation system communications.

3.1.5 Distributed antenna system

The distributed antenna system is a wireless antenna network serving cell phone, public safety radio, and other wireless applications as discussed in Section 2.3.

During an emergency, cell phone use may increase, but this will depend on the size of the building and nature of an evacuation. However, cell phone calls will be on frequencies different from public safety radio communications. Therefore, public safety radio channels are always open.

3.2 Building emergency scenarios

The main purpose of scenarios developed for this project has been to show realistic use cases where first responders are interacting with buildings. By doing this one may see practically how building networks can support incident response, and more generally how the building participates in an incident.

Scenario development entered this project at several points. First, scenarios were developed to guide our simulation efforts. This was begun as part of our efforts to collect network traffic data from NIST networks. Through understanding NIST networks and network traffic better, and having realistic incident scenarios of emergency response to the NIST lab buildings, we then could see what simulations would be most useful. Scenarios were initially developed to show response to different types of fire incidents in the NIST Advanced Measurement Laboratory.

Other scenarios were developed in addition to the AML scenarios. The first is a police school shooter scenario, and the second an emergency medical service (EMS) response to a shopping mall. These scenarios are presented below. In addition to these two, the SAFECOM SoR Vol. 1.1 Residential Fire Scenario was revised to highlight the potential role of the high-rise building to aid response, and was submitted as part of the recommended changes to the SoR (see Section 5.3).

In each scenario the first goal is to identify public safety incident response related voice and data communications on the network, as well as other applications that might introduce traffic on the building network. A second goal is to identify how the building may participate in the incident response, and what augmentation of building networks is needed to support the envisioned role of building networks, including interfaces to the public safety networks.

3.2.1 Normal building activity scenario

Fire inspections, pre-planning, and building familiarization are everyday functions of a fire department. These functions require that fire fighters enter buildings to enforce fire safety codes, note building features on pre-plans, paper or electronic, and become familiar with the building and the surrounding area.

Once inside a large apartment, office, or shopping mall type building, communication to the outside is often non-existent, which may result in missed incident calls. The public safety answering point (PSAP) will receive a 9-1-1 call for an incident in which the apparatus (fire service vehicle) that is at one of these large buildings is closest and should be dispatched. However, the dispatcher will call the apparatus on the radio and receive no response. The dispatcher is then forced to send the next closest unit which may be many more minutes farther away.

When communications are lost to a unit like this, they are considered out of service. The building network could serve as a bridge from dispatch to officers, or between officers when there is no local building emergency.

3.2.1.1 Scenario 1-A

Unit Alpha (4 men) enter a large building to perform a pre-planning activity. As they enter the building, they inform dispatch of their location and activity. Upon entering, they talk with each other using their radios, and the building network picks up these calls and establishes connections between different parts of the building. During their walk through, a call comes in and Unit Alpha is the closest. Dispatch calls them and the building network connects dispatch to the individual responders.

3.2.1.2 Scenario 1-B

Two police officers meet for lunch at a café in a large shopping mall with poor reception in the food court apart from the installed building network. As the officers are eating, a call comes in requiring their immediate action. The building network makes the connection from dispatch to officers.

3.2.1.3 Building role and network traffic evaluation

Because there is no emergency in the buildings, there is no change in the network traffic on the building network beyond the addition of the dispatch calls or fire personnel talking to each other. In each case, voice communications move across the building network. Communications from dispatch come via radio waves to a wired network node at the building exterior and move across the building network to internal rebroadcast sites.

3.2.1.4 Input to simulation work

This scenario shows a role that a building network could provide to support public safety non-emergency operations in large buildings, but it does not provide stress on the network as exists during a large emergency incident. No simulation was performed with this scenario.

3.2.2 Advanced Measurement Laboratory carbon monoxide danger scenario

The AML facility consists of five lab buildings, two of which are entirely underground. In this scenario an explosion occurs in the lowest level of the AML, but there is no ensuing fire. The explosion renders two people unconscious, and a valve on a bottle of carbon monoxide (CO) malfunctions, releasing gas in the room. The explosion has activated a smoke detector in the room, which activates the building fire alarm system.

The building fire alarm system notifies the Public Safety Answering Point (PSAP) of the smoke detector activation. The PSAP dispatches Engine 1. By the time the fire apparatus is on the road, enough CO has leaked out to activate the CO detector that is in the room. The fire panel notifies the PSAP and the dispatcher sees this notice on his monitor. The officer on the fire apparatus receives the same information. The assignment is upgraded with more fire apparatus, an ambulance, and a battalion chief. Engine 2 and 3, Ladder 1, Ambulance 1, Battalion 1 are sent.

Within three minutes Engine 1 arrives on the scene. Before getting out of his vehicle, the officer notes from the information on his computer that the only smoke detector that has activated is the original one, and that the CO level is highly elevated in that room. Another CO detector has activated in an adjacent room. The sprinkler system has not activated. The officer pulls up a live video feed from a camera on the incident floor and sees that there appears to be some damage. He also sees one of the unconscious individuals. At this point it is clear that there is no fire and that a rescue operation is required into rooms with dangerous CO levels.

The officer notifies the battalion chief that the Engine 1 crew is going in to investigate the room with the unconscious individuals. A crew of three enters the building and proceeds down the stairs. The team position and health are being monitored by the responding chief and also the communications center. Once inside the building communications with the PSAP and the Battalion Chief becomes difficult and the building network automatically takes over by providing a path for voice and data communications to be transmitted outside the building. The hand held gas detector that they carry is monitoring the atmosphere and sends the CO concentration reading to the PSAP and the responding Battalion Chief.

Battalion Chief 1 arrives on the scene and sets up his command post. Noting that there are an unknown number of individuals that may have CO poisoning he calls for additional EMS units. When the first EMS unit, Ambulance 1, arrives on the scene, they check the availability of hospitals that can treat CO poisoning, how many patients each hospital can treat, and the availability of transport helicopters, since these specialty treatment centers are 40 km away.

Chief 1 checks his incident command screen and notes that the elevators appear to be OK and notifies the crews that they can use the elevators to move the patients from the basement to the main floor. The chief gives the crew directions to the best elevators to use. Engine 2, 3, and Ladder 1 crews are sent in to assist in removing the unconscious individuals.

Fire fighters from Engine 1 find one of the two individuals, and move him to an elevator. While in the elevator the crew attaches equipment to the person that monitors his electrocardiogram and other vital signs and sends this information via the building network to ambulance 1, and this information is also transmitted to the local control hospital. At the same time, one of the fire fighters is transmitting his assessment of the patient to the ambulance. Using his radio the officer of engine 1 reports to arriving crews on the conditions that he has encountered in the basement. These transmissions use the building network. The Engine 2 crew finds the second victim and removes him.

Chief 1 requests an update from engine 3 and ladder 1 crews which are still in the building. The officer of ladder 1 uses his helmet mounted camera to send back video through the building network to the Chief of the incident scene. This video continues as the Ladder 1 crew completes the search for other injured individuals.

The ambulance crew monitoring the health status of the fire fighters in the building notices that one of engine 3 fire fighters is showing signs of distress and notifies Chief 1 who orders him out of the building.

The Chief checks his incident command screen and determines that he can start to ventilate the basement area by placing the HVAC system in exhaust and does so. He also notes that the fire fighters that are still in the building have been inside approximately 20 minutes and are nearing the end of their useful air supply and orders them out of the building.

3.2.2.1 Building role and network traffic evaluation

In this scenario, approximately two crews of three people are in the building at any one time. The officers of each crew are in voice contact with the chief outside the building. There are also communications between responders in the building. These voice communications cannot reach outside the building without some assistance from the building. In addition to voice communications, there are two sources of video in use. The first video source is the fixed building security camera. Images from this camera would be most useful if video images were high resolution (to allow far field view), although a very low frame rate is fine. The other video source is a helmet camera on one of the officers. This camera data could be lower resolution and should be faster frame rate since the officer is moving.

The helmet video and voice communications originate on the officer Personal Area Network (PAN) and are relayed to the IAN via the officer's radio. Other communications originating in the PAN include health sensor readings and handheld CO sensor.

In addition to the building security camera video stream, the scenario mentions other building sensor readings (CO, fire) that will be collected at the building information system interface and moved out to the public safety network. The scenario also mentions localizing team position. This will likely be accomplished with the help of other building referenced sensors, or perhaps antenna localization within the building and this location intelligence is communicated via the building information server to the chief.

One other application suggested in the scenario is that of the chief commanding the HVAC system to vent the incident floor. In this case the information flow is reversed and the building information system receives a command that is passed over the building network to the HVAC system controller.

3.2.2.2 Input to simulation work

This scenario provides another perspective on the potential uses of a fixed building network, as well as building information system, to support emergency incident communications and response. However, the maximum load on the network in this scenario consisted mainly of a single video signal. This also was not simulated.

3.2.3 Advanced measurement laboratory large fire scenario

An incident occurs in the lowest level of the AML. An industrial accident with an explosion causes a rupture and ignition of a gas line. Several people have been rendered unconscious. A smoke/heat detector in the room activates and sends a signal to the Public Safety Answering Point, someone close to the incident pulls a fire alarm, several people also report the incident via telephone and cell phones. The dispatcher/dispatchers who answer the calls decide that this is a serious incident with known casualties. They recommend that a "Full Response" be sent. The dispatch consists of the following:

5 engines, identified as Engines 1,2,3,4,and 5	4 personnel each	20
2 ladders, identified as Ladders 1 and 2	4 personnel each	8
1 rescue squad, identified as Rescue Squad 1	4 personnel	4
2 Battalion Chiefs, identified as Battalion Chiefs 1 and 2	1 personnel each	2
2 EMS units, identified as Medics 1 and 2	2 personnel each	4
Hazardous Materials Team, identified as HazMat Team	5	5
	Total	43

Additional smoke/heat detectors are activated in the hallway outside the room of origin and in adjoining rooms. Smoke alarms in the ventilation system are activated and the smoke control system activates, shutting off air supply to the fire floor while venting that floor, and simultaneously pressurizing the floor above the fire floor. Elevators are still operational with no smoke or heat alarms. Lights are turned on everywhere. Doors are unlocked. This building system status information is collected and sent to the PSAP by the building information system. The public safety communications distributed antenna system is enabled (due to RF restrictions in the AML underground building, all radio transmissions are normally off).

Before any apparatus arrives on the scene the dispatcher sends an additional working fire dispatch:

2 engines, identified as Engines 6 and 7	4 personnel each	8
1 Battalion Chief, identified as Battalion Chief 3	1 personnel	1
1 Safety Chief, identified as Safety Chief	1 personnel	1
1 EMS unit, identified as Medic 3	1 personnel	1
	Total	11

The units start arriving on the scene and take their assigned positions:

Engine 1	To the front of the AML, lays out from a nearby hydrant, attaches hoses to the exterior standpipe/sprinkler connection. The crew proceeds into the building and takes stairs to the floor where incident has occurred.
Engine 2	Positions at hydrant where Engine 1 has laid out. The crew enters the building and takes stairs to the floor where the incident occurred and backs up Engine 1.
Engine 3	Positions at the rear of the AML, lays out from a nearby hydrant, attaches hoses to the exterior standpipe/sprinkler connection. The crew proceeds into the building and takes stairs to the floor above where the incident has occurred.
Engine 4	Positioned at the hydrant where Engine 3 has laid out. The crew enters the building and takes the stairs to the floor above where the incident has occurred and backs up Engine 3.
Engine 5	Stands by in reserve.
Ladder 1	To the front of the AML. The crew enters building and assists crew from Engine 1 and goes to the floor where the incident has occurred
Ladder 2	To the rear of the AML. The crew enters building and assists crew from Engine 3 and goes to the floor above where the incident has occurred.
Rescue Squad 1	To the front of the AML, and crew enters building and proceeds to the floor where the incident has occurred and to the floor above the incident and attempts to shut off utilities.

Battalion Chief 1 arrives on the scene and sets up the incident command post in front of the building. He instructs Battalion Chief 2 to take care of the interior operations within the building. Battalion Chief 2 sets up a separate command post for interior operations. Chief 2 quickly checks out status of building systems and fire via his incident command screen: elevators, water availability and sprinklers, lighting and door security, smoke control and fire development and attack plans using the fire modeling decision support tool. He is keeping track of the location of all personnel in the building.

Battalion Chief 2 is also in constant verbal contact with the officers of each unit as they make their way through the building. He instructs Engine 5 now to enter the building and start searching floors for any individuals who have not evacuated. There are now 22 emergency responders in the building, each with a Public Service Communications Device. Each responder is sending out biometric information, has location tracked, and has video/thermal imagining captured and available to the incident commander upon request. In addition they could talk to each other or the command post at any time.

Battalion Chief 2 assesses the situation from the building and requests more units. He contacts the incident commander who in turn requests a 2nd alarm assignment be sent. This assignment consists of the following.

4 engines, identified as Engines 8, 9, 10, and 11	4 personnel each	16
2 ladders, identified as Ladders 3 and 4	4 personnel each	8
1 rescue squad, identified as Rescue Squad 2	4 personnel	4
1 Battalion Chiefs, identified as Battalion Chief 3	1 personnel each	1
1 Mask support unit, identified as mask support 1	1 personnel each	1
1 rehabilitation unit	1 personnel each	1
1 command bus	1 personnel each	1
2 EMS units identified as Medics 4 and 5	2 personnel each	4
	Total	36

While making their way to the fire area, Engine 1 and 2 responders have come across several unconscious persons. They radio this information to Battalion Chief 2. Since the responders now have to remove the unconscious persons, Engines 1 and 2 cannot attack the fire. They also radio this information to Battalion Chief 2. Battalion Chief 2 now orders Engines 6 and 7 to the basement to assist Engines 1 and 2. This brings the number of responders in the building to 40. Chief 2 now orders Medics 1, 2 and 3 to the lobby to start a triage area, giving a total of 46 responders in the building.

When the second alarm assignment arrives on the scene, all the engines, ladders, and rescue squads are sent into the building to assist in removing unconscious victims, searching other areas of the building, and attempting to extinguish the fire. The total number of responders in the building now would be 68. All of them would have Public Service Communication Devices (radios) tracked by the building network (or some other locator technology) and sending out biometric information. Each of them would be able to talk at any time and able to send video/thermal imagining out of the building at any time.

3.2.3.1 Building role and network traffic evaluation
This scenario demonstrates the large number of responders that may be inside a building in response to a major fire event, and includes fire fighters as well as rescue crews and emergency medical crews. Incident commanders will want to monitor everyone in the building, both health status as well as voice communications. Health status data

communications are low bandwidth as are voice communications. The battalion chiefs identified in this scenario cannot, practically, talk with more than one person or group at a time, although they may participate in multiple talk groups as needed. The largest component of network traffic is again the video connections. An incident commander may desire to monitor several fire fighter video signals simultaneously. Another chief may want to monitor rescue crew video from a helmet camera or perhaps a security camera.

3.2.3.2 Input to simulation work

This scenario was used for the simulations as a guide for the number and locations of emergency response personnel that might be communicating via a small number of network access points on the fire floor and the staging area on the floor above.

3.2.4 School shooter scenario

At 9:00 a.m., a student with multiple weapons enters the school and proceeds to walk the halls and enter classrooms and administrative spaces, dropping explosives and shooting at random to kill. School personnel call police who arrive within minutes and call in outside support including special weapons and tactical (SWAT).

The incident commander, who is the local police chief, has the school floor plans preloaded on his mobile data computer, and he subscribes before arrival to building alerts from the building information server. His main focus initially is gathering intelligence about what is going on inside. Who is in there? How many? What are they doing and where? While others interview those who have escaped, the chief pulls up the interior security camera feeds in an attempt to locate the gunman. After the first explosions he receives smoke alarm alerts that pinpoint the smoke on the floor plan. The cameras reveal the damage but do not show the gunman. The students have already been told over the public address to block their classroom doors and stay hidden.

Now, at 9:15am, the first SWAT teams enter. Observers outside watch windows and doors and pass intelligence to the chief. An intelligence officer now monitors the building information and reports to the chief. Since the school is under lockdown, all door openings are significant and all door open alerts and motion sensor activations can be used to identify locations, either of the SWAT team or gunman. An officer in the school office also can listen in on any audio from the classrooms over the public announcement system. He relays information to the chief.

With direction from the chief, the first SWAT team consisting of a group of eight slowly works its way to the gunman. At 9:25 a second larger SWAT team enters and proceeds to support the first team as well as go to areas that the shooter has been to and left, in order to check on students and identify victims. At this point there are approximately 20 SWAT team members in the building and many more police, EMS, and fire outside the building. Some of the SWAT team members could have bio-sensors and cameras that would be valuable in monitoring what is happening in the building and giving the chief additional tools for decision making.

At 9:30 a brief gun battle erupts ending when the gunman is wounded and apprehended. At this point medical teams are dispatched into the building to care for victims, and fire units enter the building to address burning debris.

3.2.4.1 Building role and network traffic evaluation

This scenario demonstrates the building role in supporting emergency responder communications as well as in providing situational intelligence for a police-focused building incident We see that video and fire and security system sensor data are all potentially important in such an incident.

3.2.4.2 Input to simulation work

This scenario was not used for simulations.

3.2.5 Emergency medical service to mall scenario

In this scenario there is a 9-1-1 call from somewhere in a large mall with only a record of a store the disabled person is near. The dispatcher sends the nearest ambulance to respond to the mall. The dispatcher then pulls up a floor plan of the mall to identify the store location on the floor plan and relays to the ambulance driver the nearest entrance information. The EMS crew also pulls up the floor plan and finds the store and upon arrival proceed to find and treat the disabled individual.

This scenario is very simple and shows only the need to help the EMS crew associate a store name that arrives in a phone call with the physical location and nearest entrance at the mall. The mall may or may not have a system to supplement radio coverage. Building information from building sensors does not play a role.

3.2.6 Scenario summary

The scenarios demonstrate the role that buildings might play in emergency building incident response. The only significant loads on the network are video signals, which include both responder helmet cameras as well as building fixed security cameras. Besides video, other important but relatively low-bandwidth traffic on the network includes responder voice and health data, as well building control system sensor data, including fire alarms, smoke control, and security system sensors.

The AML major fire scenario helped set the boundaries for simulations. All the scenarios are also useful for establishing how different buildings might participate in different incidents, including understanding how building sensor data can be used to improve emergency response.

3.3 Public safety requirements of building network

After having discussed existing building network traffic in Section 3.1, and scenarios that show what additional traffic might cross a building network in an emergency in Section 3.2, this section presents the demands that the public safety network places on the building network (or DAS) to provide mission critical voice, video, and data communications.

The demands of the public safety network on the building network include:
1. The building network (or antenna network) must meet the network performance requirements of the IAN; and
2. The IAN must be accessible inside the building so that first responders have voice communications and so that incident commanders and other personnel outside the building have access to PAN-source data including first responder location and health data as well as video.

3.3.1 Incident area network performance requirements
The Statement of Requirements v 1.2 [SAFECOM, 2006] gives qualitative network performance requirements for the IAN. The IAN must:
1. be ad-hoc and dynamic to support mobility of network members;
2. be scalable to size with the incident; and
3. provide broadband bandwidth for data and video.

These requirements apply also to the building if it is to participate in the IAN. Requirement (1) above says that there must be an uninterrupted connection as a first responder radio moves from a connection to an outside access point to an indoor access point, and that the radio connects to the indoor access point based on strength of signal. Ideally, the connection via the building is transparent, simply extending the range of the signal to allow the IAN to extend into the building. Requirement (2) above says that full building coverage needs to be provided and requirement (3) says that the building network must provide sufficient bandwidth to meet the demands of the scenarios presented earlier, ideally exceeding the bandwidth available elsewhere on the IAN.

The Statement of Requirements v 2.0 Quantitative Requirements provides some guidelines for performance of communications across a building network. According to the SoR, acceptable mission critical voice quality will ideally have a total mouth to ear delay of less than 150 ms and packet loss ratio of less than 5 %. Because the delay across the building network is only one component of end to end delay, the acceptable building network delay should be on the order of 10 ms. Acceptable video packet loss ratio requirements are more stringent, with allowable loss ratios of only 0.5 % without error concealment, and 1 % with error concealment. Video delay requirements are given as maximum 1 second. The simulations (Section 4) help to determine the ability of a building network to meet these requirements.

3.3.2 Connecting incident area network to the in-building personal area network
There are a number of requirements related to the PAN connection to the building network:
1. The signal must be available in all areas of the building. This was discussed in the section on use of DAS as a neutral host solution (2.3.3).

2. If discrete network access points are used to connect the PAN to the LAN, then mobility must be provided to allow movement and transfer of radio from one access point to another within the building, as well as smooth transition from exterior to interior. Study of mobility protocols is not part of the scope for this project.
3. The building network components must be robust and reliable so that the network is not unavailable at time of incident due to improper maintenance or failure, or lost during the incident due to physical damage. This point was addressed in the building network evaluation Section 2.2.

3.4 Provision of building-source data to public safety users

Public safety officials and first responders have agreed that building-source data, whether floor plans or real-time sensor data, can be mission critical to incident response. The building network has to deliver this information to emergency responders.

In the OLES sponsored Phase I research project [Holmberg, et.al., 2006], NIST addressed building information requirements, and the Phase I report presents those information requirements, which have been incorporated in the NEMA SB30 standard. Summarizing that work, law enforcement responders would most like to have access to information about the location of individuals in the building, particularly video surveillance camera feeds. For the fire service, building systems can provide responders with data about the location, size, and progress of a fire, in addition to information about fire equipment locations and potential hazards. At some point in the future, the building may participate in the goal of providing first responder location information.

The question is how to deliver this mission-critical building data to emergency responders. When considering the interface to building source data, the goal is to have a standard network interface that provides needed information to meet public safety applications while meeting network security and connectivity requirements. This section addresses the requirements for the building data interface which is the device or collection of hardware and software that sits at the building network perimeter to interface the building information systems to outside public safety users of the building data. It is the gateway to the building network and collects, formats and presents data to the public safety user when requested.

3.4.1 Public safety users and building data needs

NIST research has identified different places where public safety access to building information is important:
1. At the central station alarm (CSA) company. Commercial building fire and security systems typically connect to an outside CSA monitoring company. It is the job of CSA to relay alarm data to the 9-1-1 center to initiate emergency response. The CSA company often helps investigate possible emergency situations by contacting building personnel, examining alarm data, and by other means. It would be helpful to CSA if they had access to a rich building data interface to allow viewing real-time data from the building.

2. At dispatch. The dispatcher needs high-level incident data. If there is a building system generated alert, this should come via CSA. There may be a role for the dispatcher to directly access a building data interface in response to a 9-1-1 call from the building.
3. Enroute. Responders enroute need access to important building information. During the enroute phase of response, the building information most needed relates to identifying the best location to enter the building, and high-level information about the incident for incident response planning. The most likely source of data for the responder enroute will be text (audio) alerts passed on by dispatch to responding units.
4. On site at incident command. The incident commander (IC) will need access to all available building system data. The IC needs to be able to "see into" the building and monitor incident progress, viewing fire, security, and other system data. The IC should be able to implement some control measures such as commanding the smoke control system.

These public safety users building data requirements use cases indicate two classes of building data. The first is an alert with high-level incident data. The second is full access to building systems to allow examining the incident in real time. This demonstrates the requirement of the building systems interface to serve both classes of data—providing alerts on an ongoing basis in addition to allowing authorized users to access (query) the building system for more detailed information.

There are additional requirements of the building interface that may be gleaned by examining response scenarios from the previous section. The following subsections review the Section 3.2 scenarios to address building interface requirements.

3.4.2 Scenario requirements analysis

3.4.2.1 Large fire scenario

The analysis presented here refers to the AML fire scenario in Section 3.2.3. During the fire scenario: who and what are the users, applications, environment, devices, and communication networks involved in moving data from a building to an end user? The fire involves dispatch of responders and time enroute to the incident. After arrival, an incident commander has access to a mobile data computer that could be used to present real-time fire information.

- *Users*: emergency communication center dispatch; incident commander (IC); police chief; emergency medical service; fire house; police station; hospitals.
- *Environment*: Information is served to dispatchers in a potentially busy communications center where the focus is on high level incident data. In the fire and police response vehicles, the responders will best be served by audio information updates. On scene, the mobile data computer provides a rich visual interface. The primary on site user of building data is incident commander rather than the gloved fire fighter or officer in the hot zone. The onsite environment can also be characterized as one of information overload.

- *Applications*: size-up the fire (where, how long has it been burning, water required, etc); locate first responders; access to building; access routes to fire; building fire-fighting resources; locate occupants and communicate with them; get system information from [life safety: access control: HVAC: smoke control: elevator: lights] building systems; find building manager; locate utility shut-offs; access to surveillance camera views in and around building; locate doors, windows, and fire escapes.
- *Devices*: color graphical user interface with enough resolution for floor plan and virtual buttons to access additional information or system views.
- *Communication networks*: the building network connects the building information server to various building automation system controllers on the building side, and securely connects to multiple network clients on the public safety side. The building interface must authenticate to the public safety network and participate in that network according to local and SAFECOM next generation requirements.

To meet the requirements of the above system components, the building interface must:
For *Users*:
- be available on incident and jurisdiction area networks to support the users in all locations;
- provide data classes that meet the needs of these different users' information requests;
- allow for multiple connections, providing different information to different clients simultaneously;

For *Environment*:
- the information presentation in dispatch will need to be integrated with existing computer aided dispatch applications;
- for enroute responders, and audio or text presentation of important alerts might be most effective;
- on scene, the incident commander should have a rich graphical interface to access building information;

For *Applications*, the available building information may be grouped according to the following data classes to provide a cross-section of the building data as required by different public safety users:
- floor plan and 3-D view of building;
- fire size-up;
- first responder location;
- occupant location;
- access route to fire;
- hazards;
- fire fighting resources;
- elevator status;
- sprinkler status;
- smoke control status; and
- building manager contact.

For *Device*, sufficient resolution for floor plan display;
For *Communications network*:

- only allow secure connections to authenticated clients per SAFECOM network guidelines;
- support wireless and wired connections with enough bandwidth to move requested data;
- use BACnet communications to building automation sub-systems, which will require some additional BACnet protocol development (for elevators, maybe still some needed for lighting and physical security);

3.4.2.2 School shooter scenario

The school shooter scenario (Section 3.2.4) involves police and EMS response to a shooter situation in a school. This situation is analyzed here to look for differences in building interface requirements from the large fire scenario.

In this police response scenario, audio alerts may still be received enroute, but police arrive quickly and stake out the building. There are people to locate—occupants and intruders—such that video and occupant sensor data is likely the most important building information. Ideally, any building security camera can be linked to an icon on the incident commanders floor plan such that live video feeds can be accessed. Potentially the cameras can be controlled by the outside officers. Alerts from the physical access control system may help track people movement. It would be very helpful to have a room-level data classification of building information specific to the room where the shooter is located: door and window location, phone numbers, camera angles, lights, and occupant information, accessible/visible on the floor plan.

- *Users*: police chief, dispatch and emergency communications center, EMS, hospitals
- *Applications*: occupant and officer location, building 3D and 2D views, static information (phone, doors and windows, camera locations, ingress/egress paths), real-time information (lights on/off, door lock status, camera views, elevator status)
- *Environment*: incident command, emergency communications center
- *Device*: same
- *Network*: same

To meet the requirements of the above system components, the building interface:
For *Applications*, must have additional data objects:
- surveillance video streams along with camera locations and orientations
- room data—every piece of information tied to the intruder space and surrounding spaces;
- much stronger need for control, e.g., for shutting off power, locking doors, turning on lights, or disabling phone;
- phone system information;
- lighting system information; and
- access control system information.

3.4.2.3 Shopping mall emergency medical scenario

The analysis presented here refers to the AML fire scenario in Section 3.2.5. In this scenario there is a 9-1-1 call from somewhere in a large mall with only a record of a store the disabled person is near. The scenario requires direction of EMS enroute to get to the closest mall entrance, as well as floor plan information, phone caller information, disabled person status, and ideally an icon on the floor plan identifying location.

Additional building interface requirements: the main difficulty is associating a store name received as part of a 9-1-1 call with a location on a floor plan. There should be a data object that provides space names (in this case store names) along with the floor plan. A dispatch application could then do a search on a store name and have the location highlighted on a floor plan and automatic identification of closest mall entrance.

3.4.2.4 Scenario requirements summary

The building interface must be available on incident and jurisdiction area networks to support users in all locations. This implies that the building interface has a known address, that it is online, and that it is available via local wireless connection as well as via wired network connections (whether a private public safety network or the Internet). In general, a building incident will be reported via an alarm company or directly by phone call to dispatch. At that point a response is initiated, and at that point the information received from the alarm company or from the individual calling from the scene must be used to reference some building server network address in order to initiate the flow of building information to clients on the public safety network. It can also be expected that multiple clients will subscribe to that flow of information and request different data, thus requiring support for multiple simultaneous client connections.

The data interface used in dispatch needs to be integrated with current and future computer aided dispatch software. The user interface for enroute emergency responders should provide audio alerts and other high-level incident response information. The incident commander's graphical user interface should be high-resolution color.

The following data classifications would be useful for emergency response data requests:
- floor plan and 3-D view of building;
- fire size-up;
- first responder location;
- occupant location;
- access route to fire;
- hazards;
- fire fighting resources;
- elevator status;
- smoke control status;
- sprinkler status;
- elevator status;
- building manager contact;
- surveillance video streams along with camera locations and orientations

- room data—every piece of information tied to the intruder space and surrounding spaces;
- phone system information;
- lighting system information; and
- access control system information.

Finally, the building interface should provide an incident commander external control of certain building systems.

3.4.3 Additional network requirements

3.4.3.1 Data throughput, latency, quality

Public safety users need different types of building data at different times, and these have differing bandwidth, latency, and quality requirements. Real-time sensor data is generally low bandwidth except for surveillance video feeds. Besides video, the only other significant large size data file (potentially a few megabytes) that may be required is the static building floor plan information. Ideally, this information is updated regularly and a recent version kept on the responder computer such that no network download is required at the time of the incident. Throughput on a WiFi network could be constrained in general by the number of users at an incident scene although this would only be an issue for heavy video transmissions. The data channel on a standard Association of Public-Safety Communications Officials (APCO) Project 25 (P25) radio network is much more constrained, with only 9.6 kbps per channel—enough for voice and moving most real-time sensor data, but not enough for passing video and large building static data files.

While the building static data file download can be slow, real-time data transfer and video latencies must approach real-time (on the order of seconds) in order to be of use to emergency responders. In terms of quality, building real-time sensor data and static data must have guaranteed delivery.

3.4.3.2 Network security

Network security for building information was addressed in Section 4 of the Phase I final report [Holmberg, *et. al.*, 2006]. Section 4.5.3 of that report concludes that because the SoR requires protection of incident data for evidentiary purposes, the same requirements must be placed on building data. This at least requires that building real-time data is authenticated. In addition, there are likely to be privacy concerns requiring encryption of some building data transfers. The following guidelines were recommended for securing building information and the connection of the building network to the public safety network:
1. Rely on data authentication (digital signatures) to ensure that data have not been modified for investigation purposes;
2. Encrypt information that is identified by building owner as sensitive. If encryption is not available, then building owners can limit data made available to the public safety network;

3. Provide role-based access control to PSCD devices to ensure that the right users have access to needed data, and avoid pitfalls of individuals viewing or acting upon data when they are not authorized or trained to do so;
4. Archive all building data (static and dynamic) that are made available by the building from start of the incident until conclusion, for the purposes of post-incident investigation and training; and
5. Include building information and consider security of the building data and networks in the development process of any public safety network, and provide a secure path for moving building information from building information server to each public safety official who could benefit from that information, providing that data at the proper time in an understandable format.

3.4.3.3 Summary of additional network requirements

Additional network requirements include:
- Multi-megabyte per second data rate;
- Multi-client connection;
- Guaranteed message delivery;
- Low-latency (order of seconds);
- Wired and wireless data link interfaces on public safety side; and,
- Security as specified by SAFECOM, which will likely be industry standard encryption. All data transmissions authenticated. Encryption may be required across wireless and public networks. Only accept data requests from authenticated public safety clients.

4 Simulation

The goal of the simulation portion of this project was to examine the potential for routing first responder mission critical voice, video and data communications over building networks. To this end, the project team performed two separate simulation tasks. The first task was to model mission critical voice, video and data traffic from the IAN moving across a typical high-speed building network, with the use of IEEE 802.11g radio access points inside the building. The building network serves as a bridge from inside to outside. Tests were performed to examine the number of users and bandwidth that the network could support, and IEEE 802.11 interface weaknesses relative to previously developed emergency response scenarios.

The second simulation effort focused on the potential for routing current generation APCO Project 25 radio communications across a building network. In this case, the concern was not bandwidth, but the issues of number of users and the effect of the building network bridge on latency. Whereas the IEEE 802.11 interface to the building network was found to provide acceptable service for mission critical voice, video and data, the building bridge inserted in the low-speed P25 communication path leads to a significant additional signal delay.

Both simulation exercises were performed using the OPNET network modeler software [Treado, S., 2007].

4.1 Incident area network across high-speed building network

This simulation exercise was performed to examine the delay and service degradation of first responder voice and video communications across a high-speed building network. The goals were to locate bottlenecks and estimate performance of the network under the scenarios developed in Section 3.2. The purpose of the scenarios was to guide efforts to scope the number of responders in the building and look at their communication needs as a realistic event progresses. The scenarios in turn provided input to the simulations.

The most informative scenario is the worst-case explosion and fire scenario (Section 3.2.3). In this scenario the number of responding units increases until there are 90 fire and medical responders on site with a high count of 68 responders in the building. Emergency responders in the building have communication devices that are transmitting health status information to incident command outside the building. Fire responders are equipped with helmet cameras, although not all video would likely be transmitted all the time. In addition, the scenario shows that large concentrations of fire responders are located on the fire floor and in the staging area on the floor above. This concentration effect requires a solution that can prevent a communication bottleneck.

The simulations modeled the building network as a high-speed IP network (Figure 4.1). The communications path extends from the emergency responder's public safety communication device (PSCD) to an IEEE 802.11g wireless access point (AP) connected to the building network. As seen in Fig. 4.1, the APs connect via a 100 Mbps network link to the floor switch which then connects via a 1 Gbps network to the main building switch. IEEE 802.11 was used for the simulations because it is under consideration as a standard for use on the Incident Area Network. Simulations analyzed the throughput and delay from the PSCDs to the main switch. Details of the simulation model are given in [Treado, S., 2007].

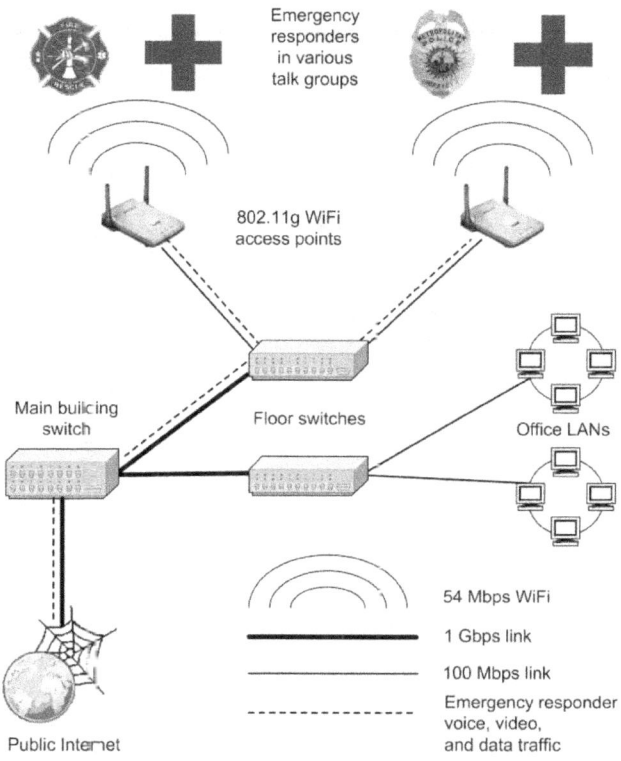

Fig. 4.1. Building network wireless access simulation OPNET model configuration.

The scenarios specify how many responders are in a given area of the building, allowing an estimate of approximately how many responders would be connected to a given AP at any time, and the types of data being transmitted. Simulated data transmissions included voice, video, and health status (signals from physiological sensors on the responder's body), of which video was by far the highest bandwidth signal. Video was modeled using both ISO/IEC Standard 14496 (MPEG-4) at 768 kbps and ISO/IEC 13818-2 (MPEG-2) at 1.52 Mbps.

Fiure 4.2 shows throughput (in percent of load) for a single AP simulation as well as a multiple AP simulation of the explosion and fire scenario. The single AP supports between 4 and 16 first responders, and multiple APs support a two phase scenario with 28 responders entering the building in the first phase and an additional 14 in the second phase. The single AP was found to support only 3.5 Mbps before dropping packets. For the multiple AP simulation, due to the nature of the scenario having a large number of responders concentrating near 1 or 2 APs on the fire floor and floor above, the APs are not able to support even a 7 Mbps load. Instead, less than 80 % of packets are successfully transmitted, and the simulation demonstrates that full video connections on only several responders is feasible with IEEE 802.11g APs. This simulation does not model the impact of internal building walls reducing throughput even more.

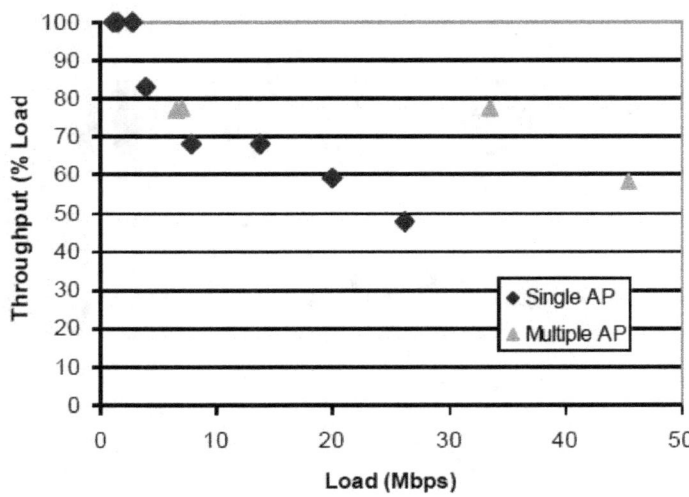

Fig. 4.2. Network global throughput for single and multiple AP scenarios, number of users ranging from 4-16 (single AP) and 28-42 (multiple AP). Throughput is low even with multiple APs due to concentration effect of first responders massing on fire floor and floor above.

Based on published SoR voice and video transmission requirements, acceptable mission critical voice quality will ideally have a total mouth to ear delay of less than 150 ms and packet loss ratio of less than 5 % [SAFECOM, 2006]. Because the delay across the building network is only one component of end to end delay, the acceptable building network delay should be on the order of 10 ms. Acceptable video packet loss ratio requirements are more stringent, with allowable loss ratios of only 0.5 % without error concealment, and 1 % with error concealment. Maximum video delay requirements are 1 s.

These simulations found the wireless AP to be the bottleneck and not transmission across the modeled building network, and demonstrated that only several video connections could be sustained with acceptable packet loss and delay performance. In actual practice, the number of video signals that an incident commander outside the building wants to view at any one time will be a limited small number. This bodes well for the potential of using existing wired building networks that may not be as high bandwidth as the simulated network. In addition, use of a wireless protocol that handles a larger number of simultaneous high-bandwidth connections (such as the IEEE 802.16-based standard also under consideration for the IAN) will likely allow more responders to connect to a single radio AP with acceptable voice, video and other data transmissions.

4.2 P25 interface simulation

For conventional P25 radio systems [Daniels Electronics, 2007], communication is comparable to that of standard two-way radios. Each talk group, a group whose members are only able to talk with each other, has a dedicated channel to use. Radios are able to directly communicate with each other, or through amplification repeaters when necessary. While this results in a simple system, it can be inefficient since it requires as many channels as there are talk groups and it does not take advantage of idle channels when the corresponding talk groups are not speaking.

Trunked communication [Desourdis, 2002] attempts to take full advantage of all available bandwidth by dynamically assigning talk groups to channels. This is an attractive approach for radio communications since frequencies for public safety communication can be in short supply in certain jurisdictions, especially for large events. In trunked P25 systems, a central base station assigns channels to talk groups that are active (i.e., one member is communicating). Talk groups are de-assigned when they return to idle. One channel is dedicated for control messages that contain specific registration and assignment information. A detailed diagram of the registration and assignment processes is shown in Fig. 4.3. Users register their radios with particular talk groups based on operational procedures, and then can hear or transmit messages from/to their talk group only. Trunked communication sacrifices some bandwidth for the transmission of control messages, but usually regains more by reclaiming unused channels when talk groups are idle.

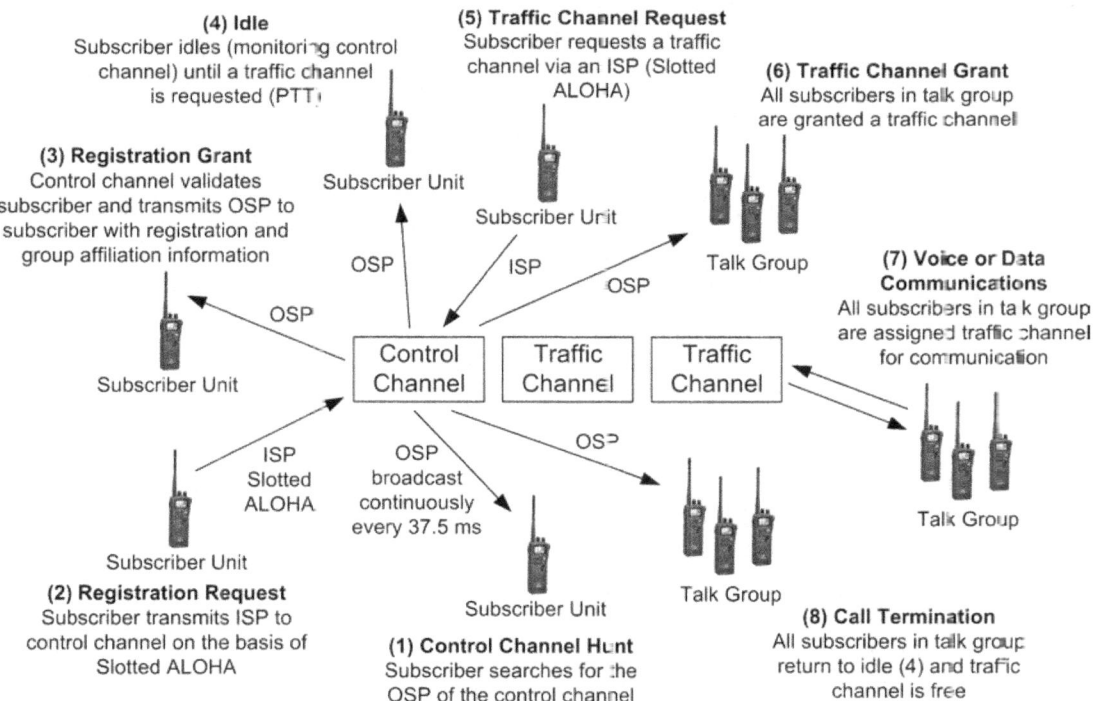

Fig. 4.3. P25 Trunking System Operation. Key: OSP-Outbound signaling packet, ISP- Inbound signaling packet, PTT-push to talk.

Trunked communication also introduces additional delay in voice communications because messages must travel from the source radio to the central base station before being sent to the rest of the talk group. Since each voice message is broken up into digital packets that require 180 ms to transmit, P25 trunked communication has a built-in delay of 360 ms (i.e., 180 ms each direction), plus any additional queuing or processing times.

The performance of a P25 radio system with and without a building area network bridge can be evaluated using network simulation computer models, such as OPNET. OPNET simulates network communication as a series of discrete events. Standard models of network components are available to simulate a wired or wireless network such as would be found in a building. Loading on the network (i.e., number of users, number of messages, etc.) can be varied to assess system performance parameters, such as delay times and the percentage of successful transmissions. Since OPNET does not have standard models for P25 radio communication, these models had to be developed for the simulations. One of the essential models that was developed represents a gateway between a P25 radio and a wired TCP/IP network. This gateway has an antenna to capture digital P25 data or voice packets and wraps them with an IP wrapper and sends them across a building network. After crossing the building network, packets are unwrapped and rebroadcast as radio packets.

In order to facilitate communication inside buildings, where signal strength from the base station may be obstructed by some kind of barrier (walls, floors, roofs, etc.), repeater systems can be established within buildings. As introduced above, P25 repeater systems can be composed of an external antenna gateway, a hub and internal antenna gateways on each floor, all interfaced with the building network (see Fig. 4.4). In a simplistic example, the hub is dedicated to transmitting P25 packets between all of the floor gateways and the external gateway. This hub would duplicate packets received from the external gateway and send each packet to all floor gateways. Packets coming from the floor gateways would be queued and sent to the external gateway. Both the external and floor gateways would perform a translation function to convert the P25 packets to and from the protocol used by the building network.

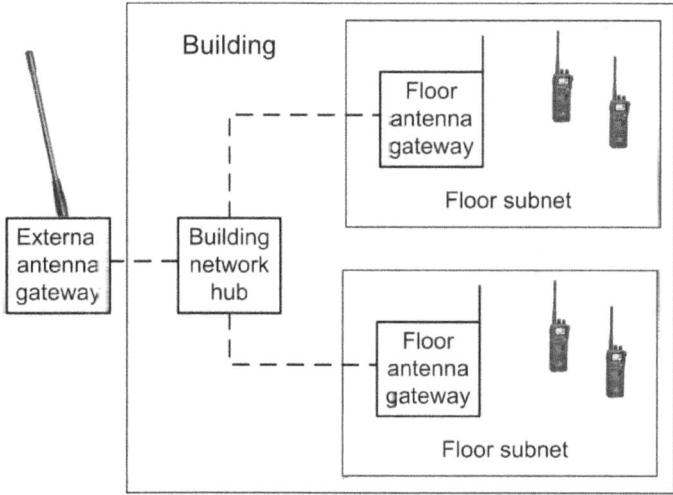

Fig. 4.4. Simple building repeater system. The antenna gateways move radio communication packets onto the wired network, with packets routed by the hub. The floor antenna gateways and radios communicate with each other on the floor level.

Rather than using a proprietary network, an existing high-speed TCP/IP building network was used to bridge between external and internal antennas, for cost effectiveness and availability reasons. This type of network adds more complexity in the programming of the antenna gateways (both external and floor) since each device would require an IP address to identify itself on the network. In addition, packets have to be converted to and from the Ethernet format and have the appropriate destinations indicated to traverse the network. However, the repeater system retains the same basic functionality—the external gateway still sends packets coming from outside airwaves to every floor gateway, and every floor gateway ultimately sends packets from floor subnets to the external gateway.

In order to evaluate the building network repeater approach, a simplified model of the building network and P25 radio system was used. The model was tailored to highlight the performance of the repeater configuration while neglecting those elements which were judged to have minimal impact on the analysis. The TCP/IP conversion process is much faster than the message transmission times, and the delay associated with that was assumed to be negligible. Also, effects due to the routing procedure itself were not considered. The basic approach of the investigation was as follows:

1. Develop an OPNET model of the trunked P25 standard;
2. Develop an OPNET model of P25 building network-based repeaters;
3. Evaluate the performance of trunked P25 communication in a conventional setting without barriers and the need of repeaters (when radios are within range of the base station);
4. Evaluate the performance of trunked P25 communication with simple building network-based repeaters (to bridge signal to radios indoors); and
5. Compare the relative performance of the two configurations.

4.2.1 P25 communication scenarios

To analyze the effect of building network-based repeaters, three main scenarios were evaluated. The first baseline scenario involves talk groups communicating directly with a central base station. It is referred to as the "outside" scenario, since it corresponds to unshielded radio communication such as would usually be the case when emergency responders are located outside of a building. The second scenario, referred to as the "building" scenario, corresponds to having emergency responders distributed throughout a building which is preventing direct radio communication. Under this scenario, the radios are divided amongst floor sub networks (based on unit ID). The third scenario, "outside & building," is used to observe the interaction of radios that use and do not use building network-based repeaters. Radios are divided amongst the floor sub networks within the building and the outdoors.

Each scenario is divided into 3 sub-scenarios on the basis of talk groups (2 talk groups, 6 talk groups, and 10 talk groups). Each sub scenario was simulated for 30 min. The message duration per radio is based on an assumed normal distribution with a mean of 3.5 s and a standard deviation of 0.9 s. The message generation interval is calculated per radio according to a normal distribution with a mean of 8 s and a standard deviation of 2 s. These values were selected for comparison purposes in order to illustrate performance issues, and are not necessarily representative of any particular communication pattern, although they are not unreasonable.

4.2.2 Global statistics

The following statistics were determined from simulation across multiple nodes (all values averaged per time bucket).
- Control End-to-End Delay (Radio to Base) – The delay between the time individual radios send control messages (channel requests primarily) and the time the base station receives them. Units: second.
- Voice End-to-End Delay – The delay between the time voice packets are sent by one member of a talk group and the time other members receive them. Units: second.
- Channel Acquisition Delay – The delay between the time a conversation is scheduled (requested) and when a channel assignment is received. Units: second.
- Message Ratio – Ratio of messages received to total message attempts per radio.

4.2.3 Control end-to-end delay (radio to base)

Control message delay is dependent on packet size and whether a message has to traverse the repeater system. Results for the three scenarios are shown in Fig. 4.5a for the case of 2 talk groups of 8 users each. For the "outside" scenario, delay is slightly more than the ideal time it takes to send a control packet (360 bits/9,600 bps = 37.5 ms). For the "building" scenario, the delay is nearly twice that of the outside scenario. This is due to the fact that each control message packet must be processed as a unit each time it passes through an antenna gateway. The gateway must wait for the entire packet to be received before passing it on. For the "outside & building" scenario, the delay ranges from that of the outside scenario to that of the building scenario, which is not surprising since the statistic records delays from radios both outside and inside the building. Spikes indicate collisions, which occur more frequently as the number of users increases, as seen in Fig. 4.5b for the case of 10 talk groups of 8 users each.

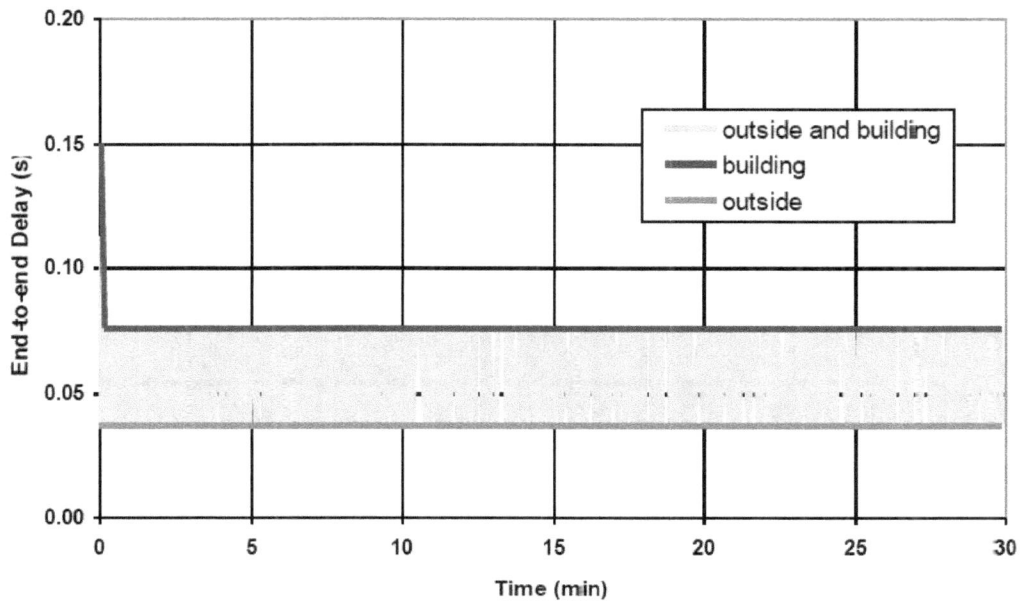

Fig. 4.5a. Control end-to-end delay for 2 groups, 8 users.

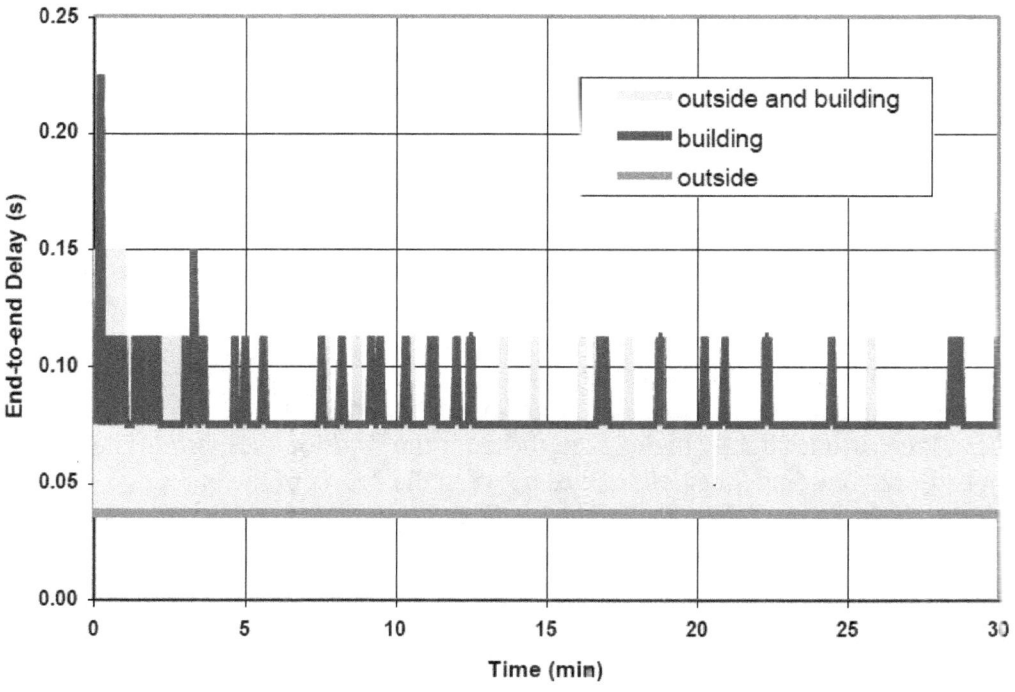

Fig. 4.5b. Control end-to-end delay for 10 groups, 8 users.

4.2.4 Voice end-to-end delay

The voice message delay is similar to the control message delay in that it is dependent on packet size and usage of the repeaters. For the "outside" scenario, the delay is slightly more than twice the time it takes to ideally transfer packets through a standard 9,600 bps stream, Fig. 4.6. For the "building" scenario (Fig. 4.6), the delay is nearly twice the outside scenario, again due to the processing requirements of the gateways. For the "outside & building" scenario, the delay ranges from that of the outside scenario to that of the building scenario. There are no observable spikes; since channel assignments ensure that only one member of each group is speaking, there are no collisions on the voice channels. The results do not change significantly with increase to 10 talk groups.

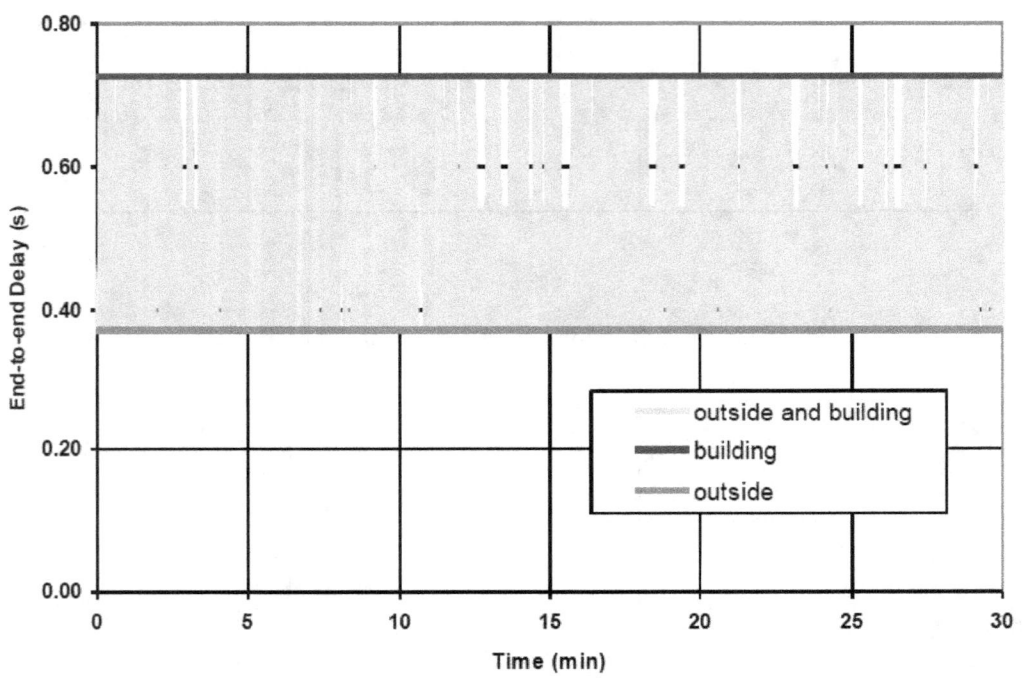

Fig. 4.6. Voice end-to-end delay for 2 groups, 8 users.

4.2.5 Channel acquisition delay

The channel acquisition delay for radios (message generation delay, Fig. 4.7) in the "building" scenario is slightly higher than those for the "outside" scenario. The delay for the "outside and building" scenario ranges between the delays of the other two scenarios. During the initial stage of the simulation, the delay is much higher than later in the simulation for all scenarios due to multiple radios attempting to register and schedule initial message transmissions simultaneously and resulting message collisions (handled by the slotted ALOHA protocol). The channel acquisition delay does not seem to change significantly when the number of groups increases, except for some initial transmissions.

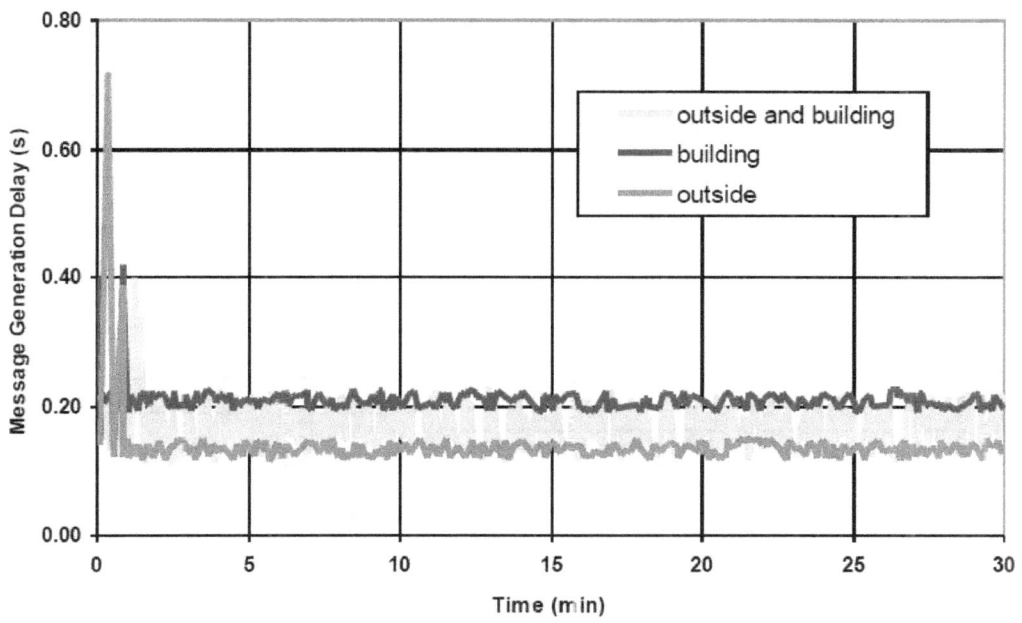

Fig. 4.7. Message generation delay for 2 groups, 8 users.

4.2.6 Message ratio

The message ratios for the three scenarios for the 2 and 10 talk group cases are shown in Figs. 4.8a and 4.8b. Message ratio is generally greater than 0.8 for all scenarios with all groups, but higher for the 2 group (Fig. 4.8a) and 6 group cases (data not shown). There were some sharp downward spikes for the 2 groups with 8 users "building" scenario (Fig. 4.8a), particularly near 17.5 min into the scenario, which is probably caused by an outlying radio(s) that is failing to establish a connection. The "building" and "outside & building" scenarios have more variability in message ratios than those in the "outside" scenarios, which suggests that the building network-based repeater system can cause some radios to experience a slightly diminished connection to the base station, while other radios are still able to experience a normal or slightly improved connection. There is a gradual improvement in message ratio with time for the 10 groups with 8 user case (Fig. 4.8b), which can be attributed to a clearing of the backlog of messages that tends to occur at the start of the scenario. Overall, the ratios for all three scenarios and progress over the 30 min scenario window appear similar except that there is higher variability in the building scenarios.

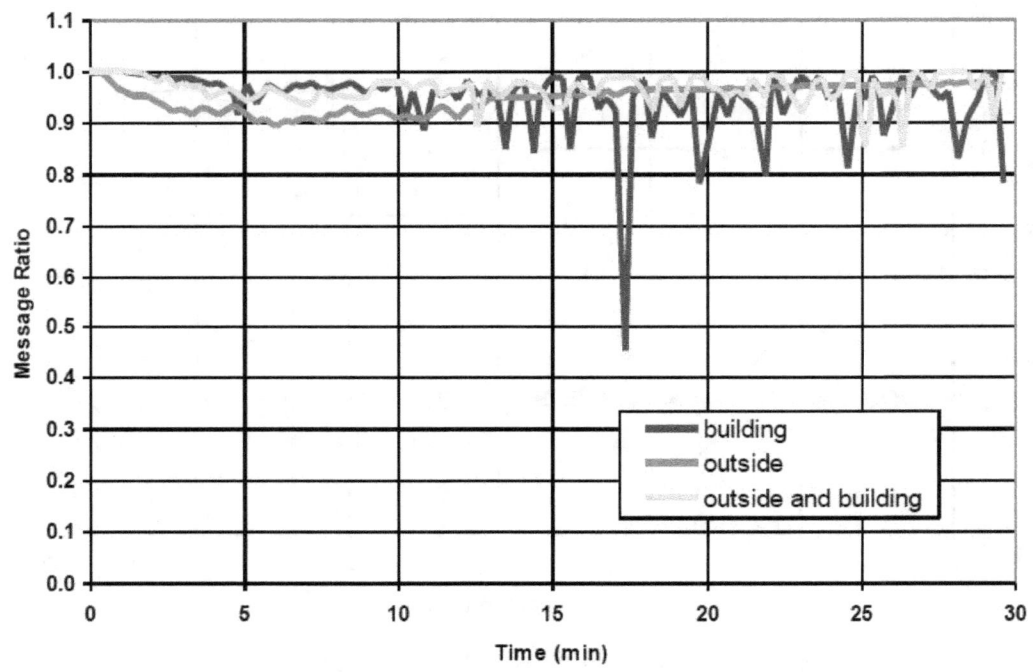

Fig. 4.8a. Message ratio for 2 groups, 8 users.

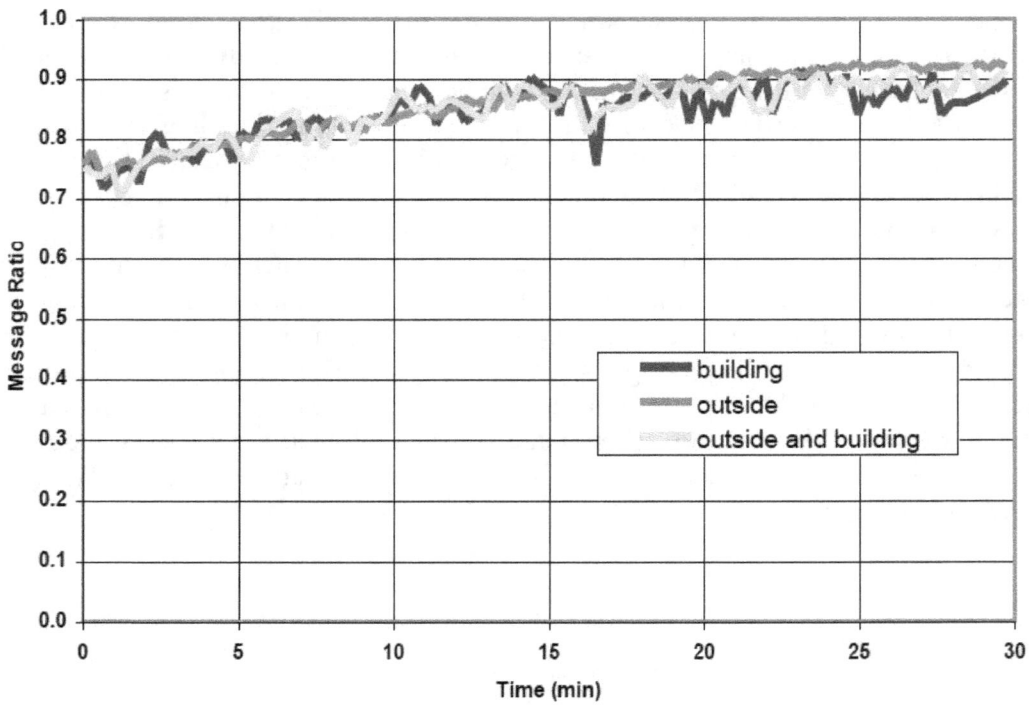

Fig. 4.8b. Message ratio for 10 groups, 8 users.

4.2.7 Analysis

Emergency responders are already concerned about the performance of P25 digital trunked radio systems [Luna, L., 2008], due to recognized potentially poor quality of voice transmissions as well as potentially significant delays in requesting and gaining channel access, especially compared to older analog radio systems where the voice is often clearer and there is no channel access delay. The SAFECOM Statement of Requirements [SAFECOM, 2006, Vol.2, Ver.1.0.] presents results of voice transmission tests where emergency responders judged acceptable voice quality and delay. The SoR states that acceptable mission critical voice quality will ideally have a total mouth to ear delay of less than 150 ms and packet loss ratio of less than 5 %. Clearly, P25 communications, with a standard voice delay of 360 ms, is already pushing beyond acceptable voice delay.

Introducing a building network bridge only makes the system delays larger and less acceptable. The building network bridge approximately doubles voice and control end to end delay since the gateways require the entire contents of a packet to be received at the gateway antenna (at a low bandwidth of 9600 bps) before converting to an IP packet and beginning transmission across the building network.

Also, given that channel acquisition delay is higher for radios using building network-based repeaters, total delay in messages, which is the sum of channel acquisition delay and voice end-to-end delay, can be nearly 1 s on average. Since these delays are transceiver based, they would be expected to occur regardless of the actual building area network protocol that bridges the gateways. In addition, there is an increased chance that radios using repeaters will timeout and drop their messages due to a combination of the delay conditions and control packet collisions (which hinder channel assignment).

However, the results show that the percentage of messages received to total message attempts may not be significantly impacted by routing across a building network bridge. The additional 180 ms delay passing the repeater system gateways produces an additional potentially harmful effect—interference. If the external base station radio signal is received within the building independent of the signal from the repeater system internal gateways, then a radio inside the building will be receiving two different packets at the same time, or essentially hearing two signals on the same frequency. This can cause the radio to fail to understand either signal. Therefore, the building network repeater system must be only used as a gap-filler, providing radio signals into building areas where there is no radio signal or only a very weak signal.

4.3 Summary

P25 simulations show that building network-based repeaters can be used to bridge trunked P25 communication between the inside and outside of buildings in a gap-filler mode. However, inserting the building network bridge into the trunked radio system approximately doubles voice and control end to end delay with total message delay approaching 1 s on average. These delays originate from the time required to receive an entire P25 control or voice packet at the radio transceiver before being able to convert to an IP packet to send across the building network. Because P25 communication delays are already greater than desired by emergency responders, inserting the building network bridge makes call setup and voice delay more pronounced for public safety radio users. Nevertheless, these results suggest that building network-based repeaters may still be a viable method of communicating inside buildings where wireless signals are otherwise blocked.

Simulations of a large fire incident where each responder has an IEEE 802.11g connection to a limited number of wireless access points demonstrated a bottleneck for high-bandwidth video signals at the access points and not across the simulated high-speed building network. The simulations demonstrate the difficulty of providing wireless access to a large number of responders concentrated in a small area within a building. Use of a wireless protocol that handles a larger number of simultaneous high-bandwidth connections will likely allow more responders to connect to a single radio AP with acceptable voice, video and other data transmissions. This bodes well for both the building network bridge implementation as well as radio connections via a DAS.

5 Analysis of building roles in emergency response

The purpose of this section is to evaluate the potential role of building networks in aiding emergency responders and public safety officials in responding to building incidents. Based on the requirements and research presented above, what roles are best suited to the building, and how can the building perform those roles? Is it reasonable for fixed building infrastructure to support incident response?

In our suggested changes to the SoR (Section 5.3 below) we present our understanding of the role that the building could have in public safety communications. In those suggested changes, we present both the potential role of building sensor data to provide situational awareness, as well as the potential for building fixed infrastructure to extend radio communications into the building. In support of those changes, we make the following assertions.

- Buildings are central to building incident response. For the fire service, buildings are involved in a high percentage of incident responses, and not an insignificant number of police responses.
- Radio coverage in buildings is a serious problem. While there are different potential solutions, the problem needs to be addressed. There can be no radio interoperability without radio connectivity.
- Large commercial buildings with modern control systems already can provide access to building system data. Public safety practitioners agree that this data is mission critical for effective and safe building incident response.
- The trend in the U.S. is toward increasing sensor proliferation and connectivity. Building data is becoming richer and more available. Building control technology is moving towards standards-based access: standard communication in the building and standard access via the Internet.
- In time, building sensor data will be available not only from commercial properties, but also residential and industrial.

The building should play an active role in building incident response. One may argue that public safety practitioners cannot rely on building owner managed systems, but the fact is that they already expect buildings to meet the building code requirements for fire equipment, fire alarm systems, and (in some places) coverage in buildings provided by building owner installed antenna systems. It is clearly reasonable to consider requiring building owners to provide for IAN communications within a facility.

Beyond radio coverage, there should be an active effort to put mission critical building information into the hands of first responders. Public safety users should have access to data from any mission critical data sensors in the incident area, including: PAN sensors, building sensors, truck sensors or field sensors. There are many data sources outside the building that may be of use in a building incident response, but none are more important than data about the building itself and what is happening inside in real-time. An ongoing effort is needed to develop a standard architecture to collect, transport, and display building information to emergency responders on scene.

Section 2 of this report documented different common building networks, of which the IT network and fire alarm system network show the most promise for carrying public safety voice and video communications. Notably, the trend for building networks is toward IP communications, with devices connected to a high-speed IT facility network. The presence of a high-speed network throughout a facility allows the potential use of this network for extending the IAN into the building.

Section 2 also reviewed the current state and potential use of in-building wireless systems for public safety communications. Distributed antenna systems are shown to be a very promising means of extending the IAN into a facility. DAS are already in use, even required by some municipal building codes, for provision of public safety radio coverage for current land mobile radio systems. Many facilities install DAS designed specifically to support public safety, while other facilities may provide a neutral-host DAS that supports multiple applications, including cell phone services and WiFi access. Trends indicate a growth in neutral-host DAS as well as in the number of codes requiring the use of DAS to support public safety in-building radio coverage.

Section 3 addressed public safety network requirements including the demands imposed by the IAN as given in the SoR, and scenario analysis to provide network requirements and building data interface requirements. Section 4 simulation results demonstrated that building IT networks can support mission critical voice and video communications.

This section examines the ability of fixed building infrastructure to meet the Section 3 requirements. Section 5.1 addresses approaches to extending the IAN into the building across existing building infrastructure. Section 5.2 examines design and implementation of a building data interface. And Section 5.3 presents our recommendations for incorporating the building into the Statement of Requirements.

5.1 Extending the incident area network

There are different potential methods for extending the IAN into a large building. In the high-rise building and large incident communications workshop [Vettori, R., *et. al.*, 2007], attendees discussed technologies in use today that help to extend communications into buildings. New York City Fire Department representatives presented their Post Radio solution which uses a pair of high-power transmitters, one of which was placed near the fire command post in the lower floors of a high-rise while the other was placed on the fire floor above. The 45 W power of the transmitters allows effective communications from lower floor to upper floors when otherwise the existing radio towers would not provide acceptable communications. This was found to be a simpler solution (cost, management) than other possibilities such as increasing the number of radio towers throughout the city.

New York City representatives also discussed the use of a cross-band repeater installed in a vehicle outside the building to boost the signal at the building. In this case, the repeater receives the signal from the municipal system on one band (say in UHF), and rebroadcasts this signal in a separate band (say VHF). This system allows for a less expensive repeater solution where the repeater does not require expensive narrow band filters that would be required if the rebroadcasting was done on a radio frequency band close to the originating frequency. The repeater boosts the signal allowing penetration into the structure.

These solutions are working in some municipalities. As noted in Section 2.3, some municipalities are relying on local code requirements for the provision of coverage in large buildings. Use of a DAS is the common tool for meeting those requirements. DAS is used to extend radio signals inside buildings, whether cell-phone, WiFi network, or public safety land mobile radio. These DAS can be designed to support a single application or multiple applications over a broad frequency range in a neutral-host arrangement (Section 2.3.2). If the IAN is instantiated in the 4.9 GHz band, or some similarly high-frequency band (relative to typical LMR frequencies), then one can expect that building structure penetration of radio waves will be diminished with more severe attenuation relative to lower frequencies. This fact increases the need to consider DAS or building network extension of the IAN into buildings, especially larger ones.

5.1.1 DAS extension of incident area network

This section addresses use of Distributed Antenna Systems to extend IAN communications into a building. Just as municipal building codes are becoming common to require owners of large buildings to provide public safety radio coverage via a DAS, codes could require an IAN interface.

5.1.1.1 Implementing incident area network on the distributed antenna system

In the case of a land-mobile radio application, the DAS within the building connects at the base of the DAS tree to a bi-directional amplifier which then sends an amplified signal to a directional antenna on the roof of the building [Section 2.3]. A similar solution can work for the IAN. As seen in Fig. 5.1, an exterior antenna near the main entrance to the building is connected to a signal booster (bi-directional amplifier, BDA) at the IAN frequencies to amplify signals and this BDA connects to the DAS to distribute the radio signal throughout the building interior. There is no delay introduced in this arrangement because the signal remains as a radio-frequency signal, distributed via coax and fiber-optic cables throughout the building. The BDA arrangement can also provide signal filtering and thus serve to reduce noise or bleed-over from neighboring frequencies.

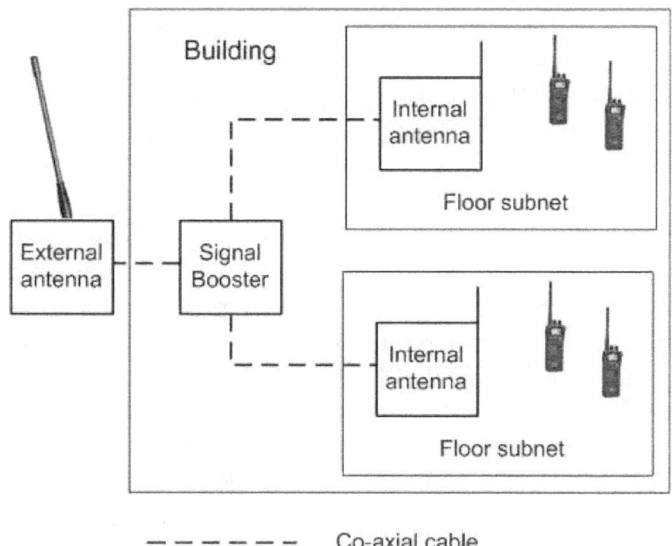

Fig. 5.1 DAS arrangement for extending IAN into building.

In this arrangement, the building is transparent to the IAN—the BDA simply boosts the signal into the building to provide coverage as needed. The only additional requirements placed on the DAS relative to current DAS solutions for public safety radio is that there be a BDA and external antenna suitable to IAN use. A different approach to this solution is that the emergency responders might carry the signal booster with them and "plug in" to the building DAS to provide needed amplification.

An alternative approach would be to have the building external IAN antenna connect to an IP access point (AP) that acts similarly to an IAN vehicle repeater. Just as the vehicle repeater links geographically separate areas of the IAN, so the building AP would link the building interior to the exterior. This solution has the disadvantage of requiring the building infrastructure components (the AP) to join the public safety network, and thus involves additional system management issues. The AP also must capture and route the data packets. The advantage of this solution is that the vehicle repeater approach also serves to provide a link to the building data interface. As discussed in Section 5.2, there needs to be a way to make mission-critical building data accessible on the IAN, and doing this via a direct connection from the building exterior antenna to the incident commander with a high-speed data connection is ideal.

5.1.1.2 Incident area network protocol suitability in building incident response

In the NIST explosion scenario, more than 60 emergency responders are in the building, and each requires a network connection to allow communications with others in the building and the incident commander outside. Because IEEE 802.11 networks are limited in number of simultaneous users per AP (typically 20 or so active users depending on packet size, and other factors [Eiger, M., 2005]), typical in-building IEEE 802.11 networks are designed with several APs per floor. If an IEEE 802.11 solution is chosen for IAN communications, then multiple interior APs would be required to connect via a wired network to one or more exterior AP. IEEE 802.16 promises better data throughput with a medium access control protocol allowing better quality of service. Potentially a single IEEE 802.16 receiver could be connected to a DAS and support 60 simultaneous connections with acceptable quality without use of a wired building network. The responder's radio (PSCD) will connect to the AP or external node with the strongest signal, switching as needed from outside network connection (at IAN vehicle AP) to interior building AP.

The SoR v1.2, Section 5.3 notes that jurisdiction area network (JAN) links can be used to tie together different parts of an IAN, and also that when the IAN link is unavailable a PSCD can connect to the IAN via a JAN interface. This makes clear the possibility that an in-building wireless system designed to provide JAN access could be used for IAN access. Thus, a responder on the outside of a building could enter the building and remain connected to the IAN by a dynamically reconfigured connection via a JAN tower. This provides backward compatability potential for current in-building wireless DAS systems.

5.1.2 Building network extension of incident area network

An alternative to extending the IAN into a building via a distributed antenna system is to attach the building exterior antenna gateway to an existing wired network inside the building. The fundamental difference between this and use of a DAS with the "vehicle repeater" AP is the complexity and management of the wired network. In the case of the DAS, the AP can be dedicated to public safety use in addition to the DAS. In the case of using a facility wired network, the network may be shared by many different users and applications, with many more pieces of network hardware attached to the network (Section 2.2.2). This section addresses some of the options and issues for implementing such a solution.

The discussion of suitability of current networks in Section 2 shows that there are hurdles to be overcome before existing networks can be used for moving emergency responder communications and distributing the signals throughout the building. In the end, what is needed for transmitting voice and especially video is high-speed networks typical of an IT general purpose network with 100 Mbps or greater data rates with network availability in all spaces in the building. Such a high-speed ubiquitous facility network could also support HVAC, security, and other system network communications, as well as first responder mission critical voice and video communications.

If the use of fixed building networks is pursued, then an arrangement similar to that shown in Fig. 5.2 can be used. The main difference is that at each antenna interface to a building network segment there must be an access point that captures IP packets from the radio frequency datalink and moves them on to an Ethernet datalink. This could lead to a significant cost relative to a DAS with only a single AP required. A packet received on an external antenna must be routed across the building network segment to the in-building APs where it is rebroadcast as a radio signal.

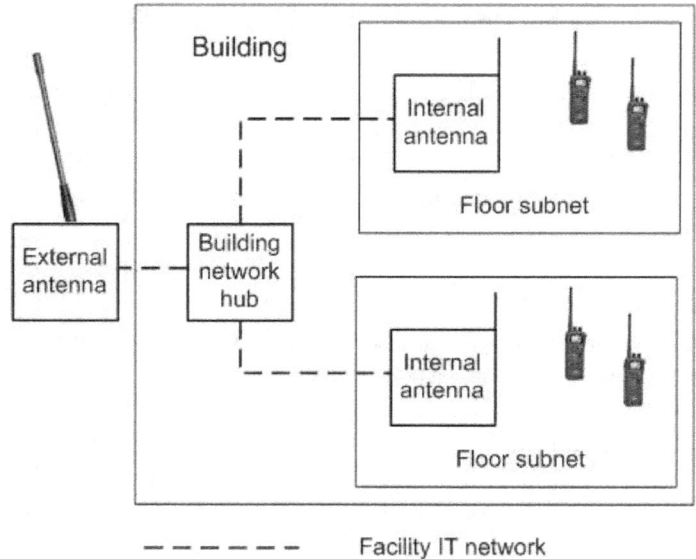

Fig. 5.2 Simple schematic of building network and antenna arrangement for extending IAN into building.

Based on the analysis of building network characteristics in Section 2 and of public safety requirements in Section 3, only two networks merit consideration for carrying IAN communications: the general purpose IT network and the fire alarm system network.

When considering moving voice and video across a building IT network (or converged facility network), the following points may be made.
1. The IT network is subject to frequent modifications that include introduction of new devices and revision of architecture. This does not lend itself well to conforming to fire code type controls that limit changes and require public safety authorities to approve changes that otherwise may disable public safety communications.
2. Along with that, control of the network is under an IT Department or delegated to third party contractors. The IT Department has many concerns and may have difficulty cooperating with public safety codes, policy, and authority. If the IT network is to serve public safety communications, then IT testing, policies, and procedures must be re-evaluated together with public safety interests in mind.
3. Security is a major issue for the IT network. Malicious attacks and error prone software leads to unreliable networks, and the constant need to patch applications and upgrade defenses (and thus adjust the network). However, security issues must be tackled as we move forward and seek to bring richer information streams out of buildings to service partners (including public safety) that need the information.
4. Nonetheless, the IT network provides a connection to a building data interface, and is already a high bandwidth IP network with security controls that can benefit public safety users.

When considering moving voice and video across a building fire alarm system network, the following points may be made:
1. The fire alarm system network is the most accessible network in areas where public safety responders need radio communications
2. The fire alarm system network is governed by code and under the authority of public safety. It is safe and reliable and robust.
3. However, the fire alarm system network is presently not capable of carrying voice and video. The same codes that make it robust also govern and limit other traffic on the network.
4. Fire alarm system network manufacturers cannot justify the cost of upgrading to high-speed IP networking and adding wireless APs without the market in place to use the APs.

5.1.3 Distributed antenna system vs. building network summary

In summary, multiple solutions exist for extending the IAN into large buildings, whether transported by emergency responders to use on-site, or whether building owner provided. Each of the solutions presented above is capable of meeting IAN requirements. If emergency responders bring a repeater or high-powered transmitters to the incident, then no interface to the building is required, but at the same time, an IAN interface to the building information server must be provided separately. For the DAS extension of the IAN, a BDA needs to be installed in the building connected to an external antenna. If the BDA is also the equivalent of an IAN vehicle node repeater, then connection to a building data interface information server is possible. In the case of a building network extension of the IAN into the building, additional wireless APs are required at each network connection point within the building to capture RF packets and move them on to Ethernet networks. A summary of key points addressing network requirements is presented below in Table 5.1.

Table 5.1 Ability of DAS and IT network to meet IAN network requirements

Network Requirement Criteria	Distributed Antenna System	Converged IT facility network
Mobility	Only one AP to cover entire building. No mobility issue	Multiple APs requires mobility protocol on building network
Reliability	Minimal changes to DAS	Frequent hardware and software and architecture changes lead to reliability issues.
Availability/ survivability	Cables/antennas can be physically protected	Network can be physically protected
Public safety oversight	DAS can be dedicated to public safety use or subject to public safety oversight	Probably not possible to submit all changes to public safety authorization, but can mandate regular testing of public safety communication

		systems
Management, testing, policy, reporting	Public safety requirements may be met via testing, policy, reporting	Public safety requirements may be met via testing, policy, reporting
Security	Network security not a concern	IT network security must be addressed for interface and sharing of building network with IAN
Multiple users	No limit on DAS, but all traffic routed through building AP (if present)	Wireless APs on each floor could be limited in number of users
Low-latency	No latency issues	No latency issues
Building data interface connection	Can provide access to building data interface if building AP used	Provides access to building data interface
Bandwidth	No bandwidth issue	Could have limitations if facility IT network is overloaded
Cost	May be no additional cost associated with using a neutral host DAS beyond adding the building AP	Significant cost to adding wireless APs throughout building, and management of public safety communications system (protocols, testing, policy, etc.)
Physical accessibility	The DAS needs to be extended to all building spaces to serve public safety	The IT network (or antennas attached to it) must be extended to all building spaces to serve public safety

5.2 Serving building-source data

Our earlier report [Holmberg, D.G., et. al., 2006] established that building-source data can be critical to effective and safe incident response, as was reviewed in Section 3.4. Section 3.4.1 discussed several phases of incident response, along with several key public safety roles that need access to information, along with how information should be presented. How can we get the building-source information into the hands of public safety officials? This section presents first the work that NIST has done so far on developing a standard for building information access, followed by details on how the users given above can get access to the building information needed at the right times.

5.2.1 Building information services and control system

The connection to the building can be via an Internet connection to a building information server (BIS). An architecture that can support this connection is under development and has been implemented in the laboratory [Vinh, A., 2007]. The system has been called the "Building Information Services and Control System" (BISACS). In short, information is collected within the building and made available on the Internet via a BISACS Base Server (BBS). The base server acts as the secure gateway to the building data—the building data interface to the public safety network (see Fig. 5.3). An outside subscriber (such as an incident commander, central station alarm company, dispatch or city planning office) may access incident data via the BBS. The BBS also notifies subscribers of new alarms with alerts sent out via the Common Alerting Protocol (CAP). These alerts go out to proxy servers that may monitor many base servers and which provide alert filtering for public safety officials who may only be interested in, for example, fire alerts versus security alerts (see Fig. 5.4). A diagram of the BISACS architecture is shown in Fig. 5.3.

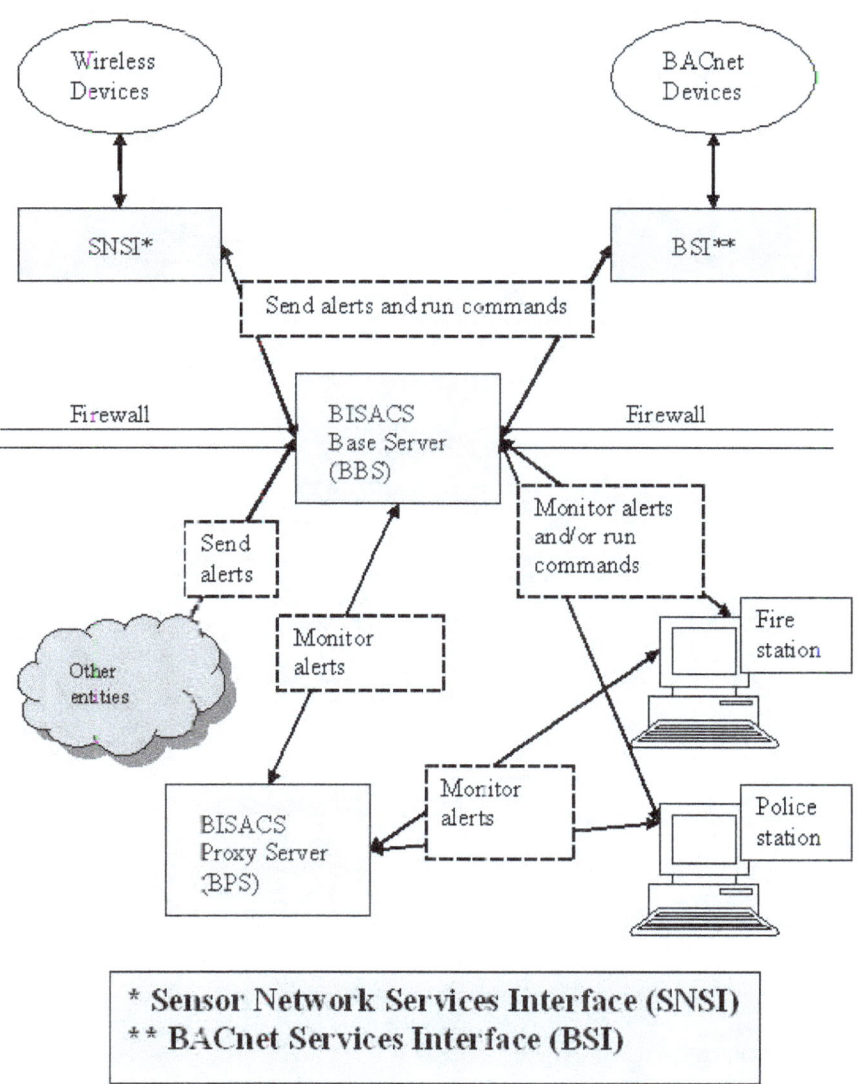

* **Sensor Network Services Interface (SNSI)**
** **BACnet Services Interface (BSI)**

Fig. 5.3 BISACS Overview Diagram

Fig. 5.4 shows how the combination of BISACS Base Servers (BBS) and BISACS Proxy Servers (BPS) can be combined to form a hierarchy of servers. This hierarchy of servers can monitor a small collection of devices at the building level with a BPS at the campus level and with higher level BPS at the local area, city, state, and possibly country levels.

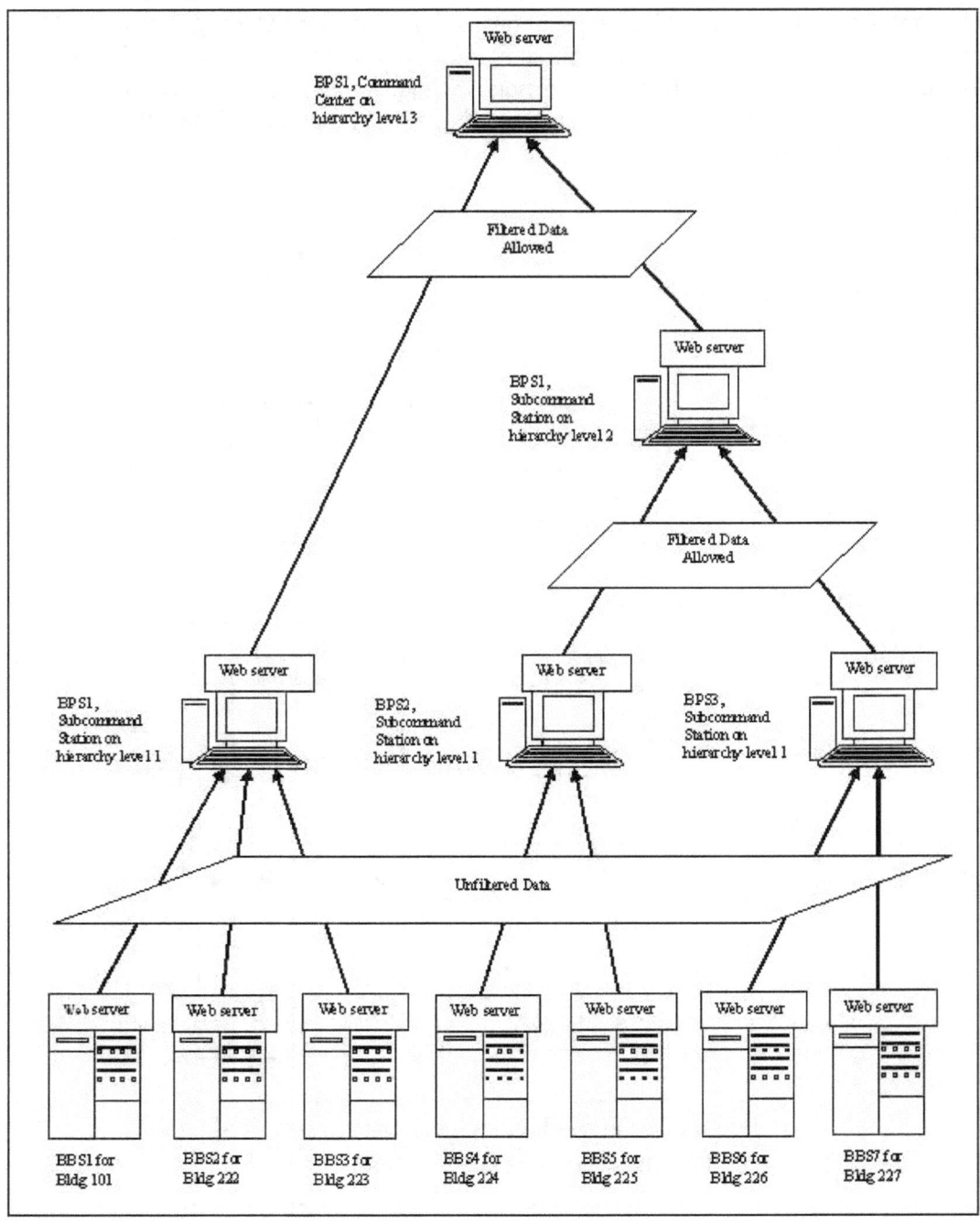

Fig. 5.4 BISACS Network Diagram

5.2.2 Getting information to the end user

Discussions with public safety system vendors have made clear that different amounts of data need to be delivered in different formats at different times. Does all this information come from a BISACS base server? How is the information transported to the public safety user? One potential arrangement for delivering needed information at the right times to the end users listed above is shown in Fig. 5.5.

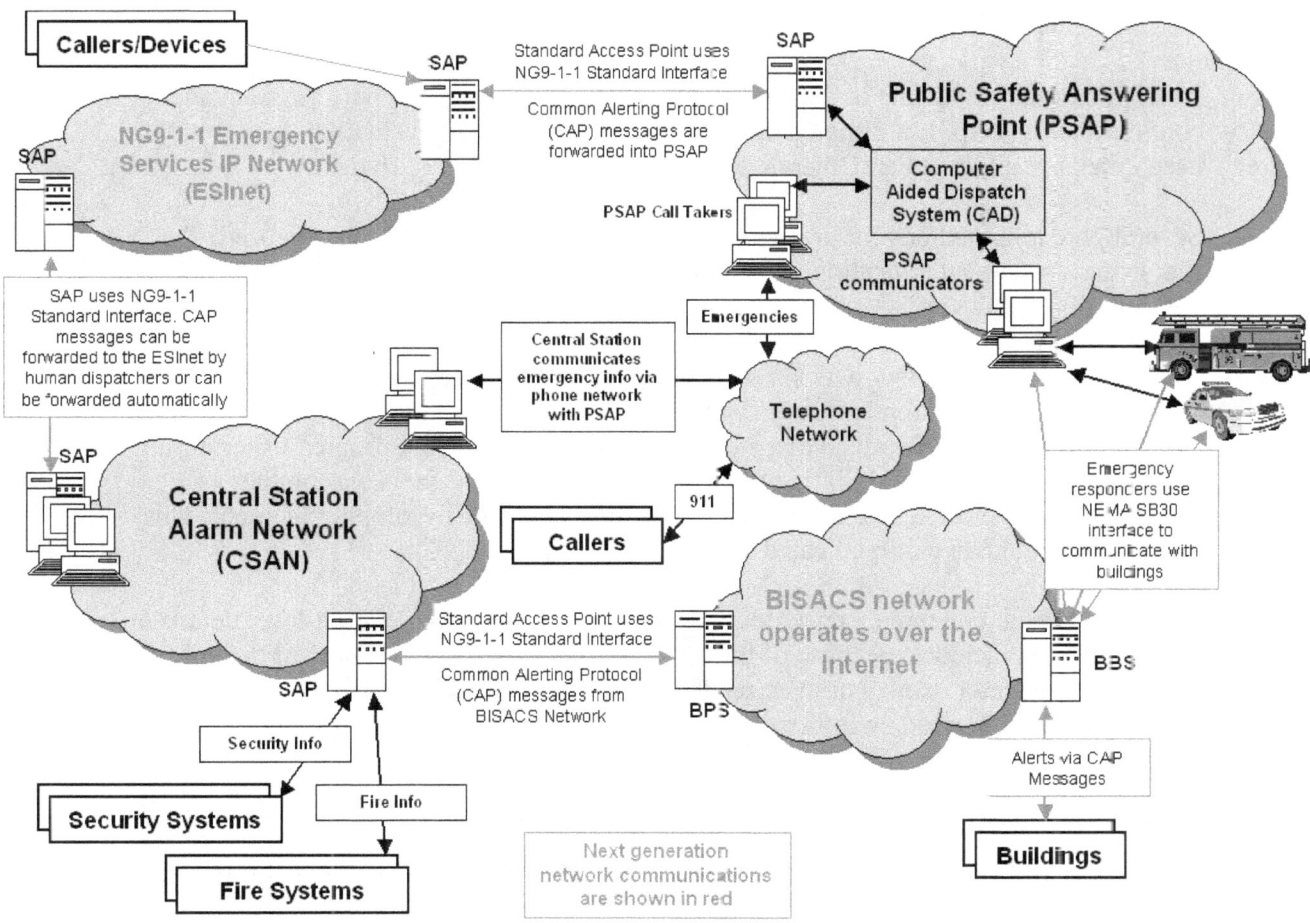

Fig. 5.5 Schematic for information delivery from building source to public safety end-users during incident progress.

The Central Station Alarm (CSA) company shown in Fig. 5.5 is the first entity to receive an alarm, although potentially any subscriber could receive the first alarm and respond to it. The role of a CSA company in the new system is not clear, but it is likely that 9-1-1 call centers will still want some intelligent agent to deal with processing and validating alarm signals and alerts prior to dispatch, and the CSA model is likely to remain valid.

The connection between CSA and the building, as shown in Fig. 5.5, implements the BISACS architecture. Therefore, the CSA connects to the building BBS via web services. The CSA could poll the BBS for alerts on a regular basis or subscribe to alerts from the BBS depending on the configuration. For smaller buildings without staff able to maintain a secure BBS, there may be alternative arrangements such as having the CSA also host the BBS. Then all subscribers to building information would access that information via the CSA-managed BBS. The connection from CSA to the emergency call center will be automated and follow next generation 9-1-1 protocols. The CSA operator/server notifies the call center of an incident in progress.

At this point the dispatcher at the emergency call center needs alert data as the incident continues to develop. If a dispatch center does its own incident evaluation, then it will need detailed building data. Otherwise high-level alerts may be sufficient.

Several potential methods for presenting data to a dispatcher have been proposed. The first is via a separate application that allows detailed alarm and other data analysis on a building floor plan, similar to what would be used at a CSA company for incident evaluation. More likely, the dispatcher could benefit from a data stream from the building that is integrated with other data sources. Instead of the dispatcher getting building address and incident summary from a caller, the building data might be automatically entered into the computer aided dispatch interface. As the incident progresses, or during initial dispatch operations, building alerts might appear as text messages on the dispatcher's screen to keep the dispatcher informed about the developing incident. This information could be communicated to others as needed.

Once dispatched, emergency responders that are enroute need to get high-level details on the progressing incident. Because of the noisy and distracted environment of the vehicle, studying a computer display or even reading text messages may be difficult. The best approach to getting incident information is via short audio messages. Responders need to know how to get to the building, where to stage their vehicles, and where and what is happening in the building so they have some idea of what they must do to tackle the incident. The high-level building incident alert information should come from dispatch, routed by the CAD system, as indicated in Fig. 5.5. These alerts can be transmitted to responding vehicles over the LMR data channel, or across the future JAN.

Once on-scene, the incident commander, with a mobile data computer, will want a high-speed connection to the building data server to allow a much richer interface to the building than would be possible with a low-speed LMR network connection. In the future network construct with IAN and JAN, it is not clear what the network data-rates will be, but it is assumed that the JAN will, like the modern LMR, be bandwidth limited, while the IAN will have a higher data rate. For this reason, we want to see a direct connection from the incident commander vehicle to the building BBS via the IAN, as shown in Fig. 5.5.

The NPSTC *Best Practices for In-Building Communications* report [Overby, S., ed., 2007] states:

> Relatively small unobtrusive 4.9 GHz access points could be placed on the outside of the building so that public safety personnel, both fire and police, could access building information as they arrive at an incident scene. The 4.9 GHz band is limited to public safety use and with proper authentication techniques could provide public safety responders a secure link over which to access information from inside the building, including video from security cameras, location of elevators, temperatures at various locations, etc. Such information could be very useful in a fire or hostage situation, as well as some terrorist event or other disaster. As of November, 2007, over 1,200 public safety agencies in the U.S. have obtained 4.9 GHz licenses.

This statement highlights the understanding by NPSTC of the potential benefit of access to building information. The scenarios presented in Section 3.2 essentially mixed the building source data applications that deliver building sensor data to the incident commander together with the mission critical voice and video applications that deliver voice, video and data from the PAN to the incident commander. The research at NIST on the BISACS architecture addresses a standard method for delivering building source data to emergency responders [Vinh, A., 2007].

Beyond the architecture for moving building incident alerts out to subscribers (the focus of the BISACS work), we must also define the mechanisms for collecting, formatting, transmitting, and presenting building information to the public safety user. This work has continued on alongside this project and a report on those efforts is in preparation [Holmberg, D.G., 2008]

Work on moving alerts to dispatch via the Next Generation 9-1-1 network, and to emergency responders on scene via a standard data interface is being pursued via cooperation with various public safety and building industry stakeholder groups and standards committees as detailed in Section 6 Stakeholder interactions.

5.3 Recommendations for SoR

One of the components of this project was to provide suggested changes to the SAFECOM Statement of Requirements that show the proper role of the building. Due to the publication schedule for the SoR Volume I, version 1.2 in fall of 2006, we submitted suggested changes in the summer of 2006, with additional more focused contribution in September of 2006. Changes were submitted to the SoR program director. This section presents the suggested changes from both those submissions along with comments for where we currently think some revision is necessary.

5.3.1 Thoughts on the place of the building in the SoR

It is our contention that the building data interface is the most important external interface during a building incident. And to the extent that the SoR is a vision document of what future emergency response looks like, the building should be included in the SoR. The SoR itself (in the section on scenarios) claims to provide "a comprehensive vision of the future of public safety communications". The potential availability of building data should be included in scenarios in the SoR.

The SoR is not in a place to require building owners to provide building data interfaces to public safety networks. And the U.S. public safety community does not want to be responsible for what they do not possess. On the other hand, the public safety community doesn't own sprinklers and fire alarms and stand pipes, or even the steps and elevators, but they know how reliable they are. They do not and cannot bring these. They can work without them. But they are in the SoR scenarios because they are familiar. Building information systems exist now, but vary widely in sophistication. They are present in a small percentage of buildings, mostly large commercial buildings serving people (office buildings, hotels, convention centers, etc.), and are unfamiliar to public safety officials.

A new high rise today will likely have a sophisticated building control system, and a fire panel with a graphical user interface that displays active fire alarms on a floor plan display. There will be elevator, lighting and HVAC systems. These systems may or may not be tied together. They could be tied to a public safety building data interface according to some federal standard or NFPA standard. Municipalities will move toward requiring these systems for buildings of a certain size and people density. And a standard interface would lead to easier adoption, public safety community acceptance, and higher system reliability with less maintenance demands on public safety.

If the building data interface is say 80 % reliable, and if the interface is available in 80 % of large buildings that public safety responders serve, then responders will use the information and come to rely on it. The SoR should point to this future and call for it. The SoR should explicitly point out the explosion in information systems and sensor density that can and will change information available in the future. So, beyond buildings, the SoR should include access to vehicle sensors in traffic accident responses such as: hazardous materials truck information (content ID, driver credentials, gas tank level, "black box" record, etc.), car information (registration info, VIN, driver details), etc.

The SoR scenarios already refer to public safety connection to traffic signals and weather systems, data access to medical bracelets, graphical information system access to fire alarm system data (presumably floor plan images from a public safety database rather than a real-time connection to the building), and even a reference to building owner provided in-building wireless IAN. So, while the SoR does not mandate building owner provision of building data and network service to public safety users, the SoR and SAFECOM should advocate for the future role of buildings in serving public safety.

5.3.2 Overview of submitted SoR change proposals

There are several main components to our suggested changes that were submitted in June 2006:

1. Section 3.3 "Residential Fire Scenario" update to include role of building network. The revised scenario is included in Appendix A.1.
2. Tables added to Section 4 "Operational Requirements of PSWC&I". In Section 4.2 "Structure Fire and Wildfire Suppression Services", a table for building data communications was added under Section 4.2.6 "Data Communications—

Interactive", and a table for building communications under Section 4.2.7 "Data Communications—Non-interactive". Similar tables were added in Section 4.4 "Law Enforcement". All four tables are given in Appendix A.2.

3. Revision of Section 5 "System of systems" that includes the role of the building in the public safety network architecture. The revised Section 5, as submitted, is included in Appendix A.3. This revision includes the building in the network architecture with links as described in the tables and is something of a grand vision for the building participation in the public safety network. This vision will not be fulfilled in the near term because too few buildings have systems in place to allow SAFECOM to plan for the building to participate in this manner. It may be appropriate now to include in the SoR a discussion of the potential role of the building in supporting emergency response with the following points made:
   a. Building owner provides IAN and JAN radio coverage as required by building codes;
   b. Building owner also provides building data interface as required;
   c. Public safety operates with available coverage and no real-time building data except where enabled;
   d. When available, emergency responders can enter the building and have building supplied access to the IAN and JAN. The building IAN gateway can additionally provide IAN access to the building data interface; and
   e. Comment on each of the links/modes given in the original June 2006 submission.

4. A new subsection added to Section 7 "Public Safety Communications Device Functional Requirements". The new subsection is titled "Building Communication System Functional Requirements" and is presented in Appendix A.4.

5. A "building information server" interface added to Section 8.2 Matrix 96, "Interface Requirements". This revised table is provided in Appendix A.5.

After reviewing the above proposed SoR changes in June 2006, the SoR Program Manager expressed concern about the lack of time to discuss changes with SoR stakeholders and requested a more focused set of less impacting changes. It was agreed that a document should be prepared that addressed potential applications in which a building could support public safety operations, with descriptions of the services, and possible updates to SoR Section 6 tables to represent the functional requirements for these applications. This document was prepared and submitted in September 2006. This document, titled "Applications for the Building Information Server to Public Safety Network Interface," is included in Appendix A.6. After submission and discussion, the SoR Program Manager agreed to the suggested SoR updates.

6 Stakeholder interactions

6.1 High-rise and large incident communications workshop

There are three components to this project. One of those components is a workshop to address "what works" today for radio interoperability at building and large incident emergency response. On June 20 and 21, 2006, NIST conducted a workshop to identify communication issues associated with high rise building incidents and to examine a variety of issues that confront public safety agencies handling large/complex incidents. The workshop report was published separately [Vettori, R., *et.al.*, 2007].

The workshop brought together police, fire, and emergency medical personnel from eight cities along with federal law enforcement personnel, manufacturers, and researchers. Presentations were given on what is working to enable communications in different areas of the United States. Breakout sessions allowed for discussion leading to the following conclusions. (1) Progress is being made in addressing the challenges of radio communications in buildings, with many solutions presented by workshop attendees. (2) For interagency communications, interoperability is less about radio patches and more about developing good standard operating procedures. (3) For large and/or complex incidents, planning, training and the use of the National Incident Management System (NIMS) are the strongest factors in determining if the incident will be mitigated successfully. (4) With large incidents, strict radio discipline is important.

6.2 NIST SoR Questionnaire

The June 2006 High-Rise and Large/Complex Incident Communications Workshop (Component 3 of this project) resulted not only in useful data on "what works" in high rise and complex incident communications, but also provided more contacts for follow-on interactions with public safety practitioners. Each of the attendees, as well as other public safety contacts, was sent a copy of the SAFECOM Statement of Requirements which included our proposed additions to the SoR, and asked a number of questions related to the content. The changes in the SoR sent to them are those presented in Appendix A. We received several responses. Two detailed responses are included at the end of Appendix B.

The questionnaire asked for input on the proposed role of the building in the IAN and of building sensors providing information to emergency responders. The responses show that emergency responders recognize the value of any available building sensor data, and the potential support of the building in providing in-building coverage by extending the IAN. Reviewers expressed that the potential for intelligent building networks and provision of sensor data would depend on the size and value of the building, given the cost of installing and maintaining such systems.

6.3 NIST Staff Presentations

NIST BFRL team members made a number of presentations to inform stakeholders about the ongoing research and results of our OLES sponsored work at NIST:

1. Bob Vettori, October 17, 2006 presentation at the Society of Fire Protection Engineers (SFPE) annual conference in Ellicott City, Maryland, "Building Tactical Information System for Public Safety Officials". He showed a shortened version of the movie of the simulated break-in and fire in Building 226 [Holmberg, D.G., et. al., 2006].

2. David Holmberg, October 31, 2006, presentation to In-Building Wireless Alliance (IBWA) and NPSTC staff at IBWA headquarters. The presentation focused on the potential for use of IBW to serve both the public safety responder communications and the building controls communications. The discussion led to further involvement with NPSTC and IBWA, as discussed below in Section 6.4.

3. David Holmberg, April, 2007 presentation "Current and Future In-Building Wireless for Public Safety" at the In-Building Wireless (IBW) Solutions Conference in Las Vegas, NV. The presentation covered requirements of IBW systems to meet public safety needs. The focus was on the possibility for building owners to provide "neutral host" IBW systems that serve not only for cell phone and IEEE 802.11 WiFi coverage, but also serve the needs of public safety. Also presented was an overview of local code requirements for public safety radio coverage. In addition, the presentation covered the SAFECOM vision for public safety networks, and the topic of potential for using IBW to locate emergency responders within buildings.

4. Bob Vettori June 5, 2007 presentation at the NFPA World Conference in Boston, "Building Tactical Information System for Public Safety Officials: Response to an Intelligent Building". This presentation had some additional material on in-building wireless and received a good response. Showed a shortened version of the movie of the break-in and fire in Building 226. Carl Peterson, NFPA liaison of a committee on fire department incident command, invited NIST to come to one of the committee meetings and give the same presentation.

5. MC Emergency Communications Center (ECC) visit, August 21, 2007. NIST team: David Holmberg, Alan Vinh, Bill Healy, Mike Galler, Bob Vettori. David presented an update on our project and the discussion focused on better understanding their needs relative to the current project direction, and on seeing how we might work together in demonstration efforts. The police representatives outnumbered the fire personnel, and were both very aware of our work and offered very good input on the project.

6. Bob Vettori, Oct 15, 2007 presentation at the Society of Fire Protection Engineers annual conference in Las Vegas, Nevada. This presentation introduced the building information model and showed a shortened version of the movie of the break-in and fire in Building 226. A member of the International Code Council came up later to discuss the need for fire/rescue and police to let the building

model programmers know what information they need. He knew about NIST and was familiar with BACnet and the Tactical Decision Aid project.

7. David Holmberg, June 12, 2008, presentation to the Alarm Industry Coordinating Council (AICC) in Washington, D.C. The AICC represents the central station alarm industry that relays building alarm information to dispatch. The presentation discussed our vision for a standard automated connection from building systems to the central station alarm company and then on to the next generation 9-1-1 network and then to dispatch. The presentation slides are included in Appendix C.

6.4 Interaction with National Public Safety Telecommunications Council

NIST BFRL staff had interactions with the National Public Safety Telecommunications Council's In-Building Working Group (IB-WG) on issues related to In-Building Wireless system interaction with building controls. A meeting was held at the In-Building Wireless Alliance headquarters in Washington, DC in October 2006 to discuss the inter-relationships of building networks with in-building wireless systems and potential for using IBW to support building control functions. NIST provided support in the preparation of the NPSTC report "Best Practices for In-building Communications" [Overby, S., ed., 2007] and input on other activities of the IB-WG. NIST staff made a presentation discussing these issues at the In-Building Wireless Solutions conference in June 2007. The work represented in these efforts is included in the Section 2.3 on In-building Wireless.

6.5 Interactions with National Electrical Manufacturers Association

The National Electrical Manufacturers Association (NEMA) SB 30 committee was established to define a standard look and feel to the fire panel's user interface to minimize confusion of an emergency responder in training on one vendor's fire panel interface and then using a different manufacturer's panel at a working incident. The SB 30 standard covers remote fire panels that allow a fire responder to view fire alarm system data remote from the building. NIST has been active in supporting this work. The work from the initial phase of the OLES funded project on information requirements for emergency responders became an important piece of the SB 30 standard. The standard also covers organization of the information on the fire panel user interface, and NIST contributed to that effort as well via our demonstrations at NIST [Holmberg, D.G., *et. al.*, 2006] and in Wilson, NC [Davis, W., *et. al.*, 2007].

However, the original scope provides no interoperability between systems, so that every building must have its own remote panel, or if enroute information is desired, then the fire service must have one of every manufacturer's remote interface devices to enable remote communications with all buildings. The committee recognized the need for a standard communications protocol to the building such that any remote client (the user

interface device) can talk to any building implementing the standard, and the fire service can have one interface device to talk with all buildings. For this reason, a new SB 30 task group has been formed to address a standard interface to the building. The standard interface will include:
- which pieces of data should be made available;
- categorization of this information;
- agreed upon sensor type list;
- agreed upon event type list;
- XML syntax and Web Services for transport of information; and
- security details.

Meetings of this task group are beginning in 2008 with the goal of having a working standard in place for the 2010 version of the NEMA SB 30 standard.

6.6 Interactions with National Emergency Number Association

The National Emergency Number Association (NENA) is the organization that oversees 9-1-1, and the more recent effort to incorporate a richer data stream into the dispatch center. This effort is called "e-911". NIST has begun to discuss with NENA the potential for integrating building data into the data stream to dispatch. We believe that the e-911 effort may be a key approach to integrating building data into dispatch while minimizing the required effort and cost.

6.7 Interactions with Alarm Industry Coordinating Council

NIST has identified the central station alarm industry as critical to moving building alerts along the path from building information system to dispatch. The Alarm Industry Coordinating Council (AICC) represents the central station alarm industry as well as other alarm industry segments. The NIST project team met with the AICC in June 2008 in order to establish a dialog to work together toward an interoperable data path for moving building alerts to dispatch. The presentation made to the AICC is provided in Appendix C.

6.8 Discussions with building system vendors and research groups

6.8.1 Honeywell

Honeywell is one of the large building system vendors, including fire and security as well as other building automation systems. Honeywell has already marketed a remote fire panel product for their building fire alarm systems One of their staff chairs the NEMA SB 30 committee mentioned above.

We have had ongoing discussions with Honeywell staff in our efforts to identify standard needs and work together on moving our OLES funded research results into the SB 30 standard. We are also working with Honeywell on cooperative research related to defining the formats for transmitting smoke modeling results. That is, we have computer models that analyze smoke and heat sensor data to identify fire and smoke locations and then process that data to determine current smoke layer and temperature danger levels and to estimate future fire and smoke propagation. Then a standard format must be defined for communicating those results to a remote client that can then display those results on a building floor plan.

6.8.2 Siemens
Siemens is another international building controls vendor. We have regular interactions with Siemens engineers related to building security system standards, and believe Siemens will be a key member in helping to define standards for police access to security system data from buildings.

6.8.3 Simplex
Simplex and GE/Edwards are important partners in access control systems. NIST has a Simplex fire alarm system and we would need a Simplex BISACS interface for a demo. We held a teleconference with Tyco/Simplex representatives in October 2006. The Tyco staff were interested in our view on how they might route emergency responder communications over the fire alarm system network using Zigbee and WiFi. The real issue seems to be cost: higher cost for wireless communications and a high-speed network when there is no established market for radio access to the fire alarm system network.

6.8.4 Motorola
We were contacted by Motorola and discussed our work on the remote user interface. Moving forward, we hope that Motorola with participate in standards activities related to communications of building information given their important place in the public safety community.

6.8.5 Johnson Controls
Johnson Controls is another major building system vendor and is a leader in integration of wireless, both wireless control networks as well as provision of in-building wireless for radio and cell communications. We have had ongoing interactions with Johnson related to building controls communications standards.

6.8.6 University of New Hampshire Interoperability Lab's Project 54
The UNH Interoperability Lab has a project (Project 54) focused on automating many of the tasks in the police cruiser, allowing an officer to use voice commands to tune the radio, turn on siren, and perform other services. Bill Lenharth, Laboratory Director, visited NIST in March 2007 to see how NIST might cooperate in research toward implementing residential fire alarm system integration into the public safety information stream. The goal is that residential fire alarm systems have an Internet connection that allows transfer of fire alarms to some regional location that can pass that data on to the police and fire services to give them early warning of working fires in residences. This is similar to our goals, but our focus currently is on large buildings.

We expect that we will be able to cooperate in future research efforts. UNH offered to work with us on software development and system demonstrations with local law enforcement when NIST is ready for that.

6.8.7 United Technologies (UT)
United Technologies is an international company with a research center that is active in many building controls related areas. We met in April 2007 to discuss research overlap. UT now owns Automated Logic and Otis Elevators, both of which are important companies in the building controls market. NIST is working now to encourage the development of an elevator system BACnet interface to allow standard communication of elevator systems into a first responder application.

6.8.8 NetTalon
NetTalon representatives met with us at NIST in June 2007. NetTalon has a product which is essentially a proprietary version of our vision—to move fire and security data from a building into the hands of emergency responders. They're interested in incorporating NIST science, whether fire modeling, or the icons work in NFPA. Their System 3000 was UL listed in February 2007, and they have systems being installed or evaluated in four locations. They mentioned ongoing live burn tests and active shooter tests and offered to cooperate with NIST in data collection for fire modeling or other purposes.

6.8.9 ISMS
We were contacted by ISMS in May 2007. ISMS is a small company with focus on "intelligent exit signs". Their goal is to tie into the building controllers and extract information about the progress of an emergency event so that they can have dynamic signs to guide building evacuation. They were very interested in any standards that would support integration of building systems and access to emergency alarm and other data. They expressed the desire to work with us in standards development.

6.8.10 Dione Systems
We were contacted by Dione Systems in November 2006. Dione Systems is a small company that aims to connect employees with real-time information about incidents in the building, giving access to information via cell phones, pagers, and desktop workstations. This requires standards for pulling data from building systems, and therefore are interested in working with us on moving the standards forward.

6.8.11 Eutech Cybernetics
Eutech Cybernetics contacted us in February 2006. Eutech Cybernetics is a small company based out of Singapore and which serves large commercial building owners with middleware for building asset management. Their iViva.works product includes ties to the building controller for collecting alerts information that can recognize building events and trigger processes that can intelligently respond to the event as for crisis management. They are very interested in standard access to building data.

6.8.12 MIJA

MIJA representatives visited NIST in April 2006. MIJA has a product that allows active monitoring of safety equipment (e.g. fire extinguishers) with a wireless connection to the fire panel. In this way, equipment can be monitored remotely and automatically. We discussed the potential use of this type of wireless connection for linking first responder PAN equipment (such as air pack) back to the fire panel as a connection to the IAN.

6.8.13 Bentley

We met with representatives from Bentley in December 2005. Bentley is a leading provider of integrated building information modeling (BIM) solutions and their geospatial division provides a complete portfolio of graphical information systems (GIS) and engineering solutions. Bentley staff came to NIST to better understand our work: how we see integrating facility models into an emergency responder display, how we see bringing dynamic sensor data into that display, and how to move the work on standards along. We found overlapping interest in communicating to stakeholders how BIM, performance-based standards, and GIS/CADD interoperability can deliver lifecycle benefits to owner/operators, public safety teams, and others who leverage those aspects of infrastructure information to support facility and campus operations.

6.9 Stakeholders summary

NIST has connected with many of the stakeholders that need to be involved in moving standards efforts forward. These stakeholders include public safety practitioners, vendors of public safety equipment and software, as well as building information system vendors. A recurring theme among the vendors is a desire to connect public safety systems to building systems and the need for standards to do that. NIST looks forward to working with these stakeholders to move standards forward.

7 Summary

The public safety community in the U.S. is moving toward an interoperable network of networks that includes coverage at building incidents via an incident area network. The challenge is providing in-building radio coverage as part of the IAN. NIST has studied the possibility of using fixed building infrastructure to link responders inside large buildings with incident command outside the building.

Wired building networks (IT, fire, mechanical, physical security), typical of those found in modern commercial buildings, were examined for suitability for carrying emergency responder communications. The IT network was found to meet IAN bandwidth, latency, and quality of service performance requirements, but implementation would be challenging due to cost, management, and political issues related to cooperative oversight shared with local public safety authorities. On the other hand, a typical fire alarm system network cannot currently handle high-bandwidth traffic, but is secure, robust, redundant, with guaranteed alarm priority, trusted by public safety officials, and available in every building space. Potentially either of these networks within a building could be designed to bridge emergency responder voice, video, and data communications from interior to exterior.

Distributed antenna systems are seen to have promise for providing coverage for current and future public safety radios, and discussion of different implementation possibilities is presented. A summary table showing the suitability of DAS and building network solutions to meet IAN network requirements is given in Section 5.1.3.

NIST simulations of a large fire incident where each responder has an IEEE 802.11g connection to a limited number of wireless access points demonstrated no limitations for transmission of mission critical voice, video, and data across the simulated high-speed building network, but with a potential bottleneck for high-bandwidth video signals at the access points due to IEEE 802.11 limitations. The simulations demonstrate the difficulty of providing wireless access to a large number of responders concentrated in a small area within a building. Additional simulations were performed to analyze the potential of building networks to transport current P25 communications. The results show that such an arrangement is feasible but introduces a significant delay due to the low-speed P25 data rate and time required to capture and subsequently rebroadcast the data packet.

It is not clear how the IAN architecture will develop, but the building should be included in the plans, along with consideration for the development of in-building communications via DAS or wired networks. In particular, there should be cooperation between SAFECOM and in-building wireless product manufacturers to develop a strategy to guide DAS development such that future DAS may support the IAN in-building coverage needs. Likewise, SAFECOM should be involved in the development of interoperability solutions that will provide building data to emergency responders.

The role of the building in emergency response is discussed in Section 5 along with NIST recommendations for inclusion of the building in the SAFECOM SoR. It is our hope that SAFECOM will pursue the full potential for building integration into the future interoperable public safety networks.

In conclusion, some recommendation for moving forward:

1. Building sensors and availability of rich building data are increasing. Plan for integration of buildings into the future interoperable public safety networks. Consider an update of the SoR with the role of the building and other data sources made a clear part of the future network vision.

2. Cooperate with the In-Building Wireless Alliance and public safety stakeholders to encourage the use of neutral host distributed antenna systems. Watch the trends toward building codes requiring these systems and consider the potential for utilizing DAS to extend the IAN.

3. Stay connected to the work in NEMA SB 30 on an interoperable building data interface so that this developing building interface can be integrated with the IAN network.

8 References

Cox, J., "Airport covered by radio-on-fibre: Passengers and staff share the infrastructure," Techworld, May 2006 article: http://www.techworld.com/features/index.cfm?featureID=2559

Daniel, J., "Signal Booster Codes and Ordinances," white paper available at: www.RFsolutions.com, 2005.

Daniels Electronics, Ltd. (2007, Jan.). P25 Radio Systems- Training Guide [Online]. Available: http://www.danelec.com/library/english/whitepapers.asp

Davis, W.D., Vettori, R.L., Reneke, P., Brassell, L., Holmberg, D.G., Kostecki, J., Kratchman, J., "Workshop on the Evaluation of a Tactical Decision Aid Display," NISTIR 7268, October 2005.

Davis, W., Holmberg, D., Reneke, P., Brassell, L., Vettori, R., "Demonstration of Real-Time Tactical Decision Aid Displays," NISTIR 7437, August 2007.

Desourdis, R. I., Jr., Smith, D. R., Speights, W. D., Dewey, R. J., DiSalvo, J. R., Emerging Public Safety Wireless Communication Systems. Artech House, 2002

Eiger, M., et al., "The Effect of Packetization on Voice Capacity in IEEE 802.11b Networks," presented at the Consumer Communications and Networking Conference, January 2005.

Holmberg, D.G., Davis, W.D., Treado, S. J., Reed, K. A., 2006, "Building Tactical Information System for Public Safety Officials," NISTIR 7314, January 2006.

Jacobsmeyer, J.M., "Proposed Municipal or Fire Code For Public Safety Indoor Wireless Coverage", Pericle Communications Company, February 2004.

Jones, W.W., Holmberg, D.G., Davis, W.D., Evans, D.D., Bushby, S.T., Reed, K.A., "Workshop to Define Information Needed by Emergency Responders during Building Emergencies," NISTIR 7193, January 2005.

Lenel Systems International, Inc., "Lenel Intelligent System Controller," http://www.lenel.com/utcfs/ws-464/Assets/OnGuard_HardwareCatalog_200706.pdf, 2007.

Luna, L., "Va. LODD highlights digital radio system issues," Mobile Radio Technology, March 28, 2008. [Online] Available: http://www.firerescue1.com/communications-interoperability/articles/393385-Va-LODD-highlights-digital-radio-system-issues/

Overby, S., ed., National Public Safety Telecommunications Council, "Best Practices for In-Building Communications," November 2007

Public Safety Wireless Network Program, "Public Safety In-Building/In-Tunnel Ordinances and Their Benefits to Interoperability Report," available online at www.safecomprogram.gov, November 2002.

SAFECOM Program, "Public Safety Statement of Requirements for Communications and Interoperability, Quantitative," Volume 2, Version 1.0, August 2006.

SAFECOM Program, "Public Safety Statement of Requirements for Communications and Interoperability," Volume 1, Version 1.2. Available: http://www.safecomprogram.gov/SAFECOM/library/technology/1258_statementof.htm, August 2006.

Treado, S. J., Holmberg, D. G., Cook, S., "Simulating the Performance of Building Area Networks as a Communication Bridge to Emergency Responders," in Proc. of OPNETWORK 2007 Conference, August 2007.

Vettori, R. L., Lawson, J. R., Davis, W. D., Holmberg, D. G., Bushby, S. T., "High-Rise and Large/Complex Incident Communications Workshop," NIST Technical Note 1479, February 2007.

Vinh, A., "Building Information Services and Control System (BISACS): Technical Documentation, Revision 1.0," NIST Interagency Report 7466, September 2007.

Virginia House of Representatives "Report of the HJ588 Task Force to the VA House," 2003, available at: http://www.vafire.com/government_affairs/House%20Joint%20Resolution%20588%20-%20November%202003v2.pdf

9 Appendices

A. NIST recommended changes to SAFECOM Statement of Requirements
B. NIST SoR Questionnaire Feedback
C. Standards for Building Information Exchange with First Responders: presentation to the Alarm Industry Coordinating Council

Appendix A: NIST recommended changes to SAFECOM Statement of Requirements

This appendix present recommended changes to the SAFECOM Statement of Requirements that were submitted to our sponsor. The changes are organized according to the relevant subsection of the SoR. Details are provided section by section below showing deletions in strikethrough font and additions as underlined text.

A.1 SoR v1.1 Section 3.3 Fire-Residential Fire Scenario, revised fire scenario

A.2 SoR v1.1 Section 4 Operational Requirements of PSWC&I, new Building Data Communications tables

A.3 SoR v1.1 Section 5 System of Systems, revised including building in network architecture

A.4 SoR v1.1 Section 7 Public Safety Communications Device Functional Requirements, new "Building Communication System Functional Requirements" subsection

A.5 SoR v1.1 Section 8.2 Network Functional Requirements, Matrix 96, additional building interface requirements

A.6 Applications for the Building Information Server to Public Safety Network Interfaces

Appendix A.1 SoR v1.1 Section 3.3 Fire-Residential Fire Scenario, revised fire scenario

(proposed additions shown as underlined text, and deletions shown as strikethrough)

3.3 Fire-Residential Fire Scenario

3.3.1 Initial Work Shift Tasks

1. Three firefighters begin their shift at the Brookside Fire District Station BFD-7. After completing their administrative check-in, they complete their biometric identity check with their PSCDs. After authenticating each firefighter, the system sets up their profiles on their PSCDs and the network, establishes the level of data access that each is authorized to have across available databases, and initiates personal tracking of each firefighter so that a record can be made of all instructions that are given to each, and the actions and responses of each firefighter. The firefighters initiate the equipment self-tests of the vests they will wear during a fire situation. The vests measure each firefighter's pulse rate, breathing rate, body temperature, outside temperature, and three-axis gyro and accelerometer data. Each vest also provides geolocation information for the wearer and measures the available air supply in the firefighter's oxygen tank. The vests have a self-contained PAN that interrogates each of the sensors and monitors. The vest codes the firefighter's information with that firefighter's ID and then transmits the data to that firefighter's PSCD.

2. The firefighters begin their check-out of the fire equipment, the fire engine, E7, and fire ladder, L7, at the station. Each apparatus has sensors to measure water pressure, water flow, water supply, fuel supply, and geolocation. Each apparatus also has its own PAN for interrogating all apparatus monitors. The apparatus codes the apparatus ID with the measured values and geolocation information for routing to the network. After successfully completing all the self-tests, the firefighters provide a digital status to the network that they have completed all initial setups and they are ready. The fire station network reports to the dispatcher, via the station's and onboard data systems, which personnel and equipment are active and available for calls. The station battalion chief follows up with a PSCD voice call with the same message. The dispatcher acknowledges that BFD-7 is active and that dispatch's Geographical Information System (GIS) and CAD system are properly receiving location and status data from the units.

3.3.2 Fire Response to a Residential Fire Call

1. At 3:17 a.m., ~~the Brookside PSAP receives a 9-1-1 call from a cab driver that the~~ <u>a smoke detector has activated in Apartment 1202 in a 20 story</u> apartment building at 725 Pine ~~is smoking and appears to be on fire.~~, <u>and the building information server (BIS) for 725 Pine transmits this information to the Brookside PSAP.</u> From the CAD display, the dispatcher finds that the BFD-7 station is available and close to the address. The dispatcher notifies BFD-7 to send E7 and L7, and to send the BFD-7 battalion chief as the fire's incident commander (IC). As E7 is leaving the fire station, firefighter F788 jumps onto the back of the vehicle. The vehicle registers that F788 has become part of the E7 crew for accountability and tracking. <u>As the fire apparatus leaves the station, the officer on E7 consults his Mobile Data Computer in the apparatus and is presented with the "En-Route" screen which shows a plan view of the building, location of fire hydrants and exterior fire department connections to the building standpipe and automatic sprinkler systems. The "En-Route" screen also</u>

gives real-time information relayed from the BIS: the location of the alarm, the closest entrance door for the fire department to use, the closest stairwell and interior standpipe connection for the fire department to use, and the location and status of the closest elevators for fire department use. The officer on L7 and Battalion Chief 7 all have the same information as does the dispatcher back at the PSAP. Before any of ~~The dispatcher simultaneously sends a digital message providing the apartment building's address.~~ the fire department apparatus arrive on the scene a second smoke detector in Apartment 1202 has activated. The BIS transmits this information to all the responding fire apparatus and the dispatch center. The dispatcher, noting that the second smoke detector in the apartment has activated, decides to upgrade the response. The dispatcher notifies another Brookside Fire Department, BFD-12, to also send an engine to the ~~fire. By 3:19 a.m., E7, L7, and the IC leave BFD-7 and report their status to the dispatcher. As the IC's command vehicle leaves the station, a nearby wireless PSCD sends the apartment's building plans and the locations of nearby fire hydrants, the building's water connections, the elevator, and the stairwells to all en route fire vehicles GISs, including the command vehicle.~~ fire, and notifies the responding units and the IC of this. The dispatcher sends a reverse 9-1-1 call message to all residents of the building, which has ~~eight~~16 apartments on each of ~~three~~20 floors. The nearest ambulance is alerted by the dispatcher to proceed to the scene. The local utility is alerted to stand by for communications with the IC at 725 Pine.

2. The E7, L7, and IC drivers view the apartment's address on the cab ~~monitor~~ mobile data computer display~~s~~, which also maps the route for the drivers; a computer-activated voice directs the drivers to the appropriate lanes and where to turn. As the fire vehicles approach traffic lights along the route, the onboard signaling system changes the lights in the emergency vehicles' favor and the geolocation system provides the vehicles' location and progress on the dispatcher's CAD display. The onboard system also interrogates the county's transportation system for road closures, blockages, train conflicts, or slow traffic conditions to route the vehicles around impediments and provide the fastest route to the fire.

3. The IC arrives on-scene at 3:22 a.m., assesses the situation by switching his mobile data computer to the "on scene" Screen. This screen shows the IC the floor plans of the building, location of utility shut offs, location of interior fire department stand pipe connections, and any other information the fire department has entered into the data base such as the location of any individuals that may need extra assistance in evacuating the building. The IC changes his view on the "On Scene" screen to show the floor plan for the 12th floor, on which the IC sees the two smoke detectors that have activated in apartment 1202. The two rooms is which the smoke detectors have activated are now shaded in ~~situation, noting that smoke and fire are visible, and alerts dispatch that 725 Pine is a working fire. The IC directs the local utility to shut off the gas to 725 Pine~~ green, which indicates that there is enough smoke in the room to make visibility difficult. As L7 and E7 arrive and move into position, all fire personnel and equipment are shown on the IC's GIS display. The system automatically sets up the tactical communications channels for the IC and the fire crews. The fire crews are able to talk continuously with each other, reporting conditions and warning of hazards. ~~Because the apartment building is not large enough to require a built-in wireless IAN for emergency services, the first fire crew into the apartment drops self-organizing wireless IAN radio bridges on each of the floors as they progress through the building.~~ Soon E12 and the assigned EMS unit arrive on site. The new personnel and equipment are automatically registered with the IAN, and their PSCDs are automatically reprogrammed to operate on the incident's PSCD radio channels and protocols.

4. As the crew from E7 and L7 enter the building the building fixed networks help track their location and send this information back to the IC. When the crews enter the elevators, their PSCDs cannot transmit to the outside of the building and the building antenna system automatically picks up the radio signals and assists in transmitting voice and other data transmissions to and from the crews that are in the building to the IC.

5. At this point a third smoke detector has been activated on the 12th floor. This time it is a smoke detector in the hallway just outside of apartment 1202. The BIS transmits this data to the IC who is notified of this through his mobile data computer. The IC switches back to his "On Site" screen and to the 12th floor plan and now notices that the room where the original smoke detector had activated is shaded in a yellow color indicating that this space may now have a toxic thermal atmosphere. The room where the second smoke detector had activated is still shaded in green and now the hall way outside of apartment 1202 is shaded in green. The IC radios the officers of E7 and L7 to advise them of the changing conditions on the fire floor.

6. Several families have already evacuated the building. As firefighters ask for their names and apartment numbers, they use the voice recognition capabilities of their PSCDs to capture the information, applying an RFID wrist strap to each resident to track their status and location. Other firefighters enter the building to guide survivors out and to rescue those who are trapped. ~~The~~ As the crew from E7 and L7 proceed to apartment 1202 the infrared (IR) cameras on the firefighter's helmets provide the IC a real-time, on-demand view of fire conditions within the building and the location of the hot spots. Additionally, the firefighters monitor the temperature of the surrounding air in their location; this information is directly available to the firefighter, as well as the IC and EMS unit on-scene. Other passive sensors, such as hazardous gas detectors, are also operating in the firefighter's PAN. ~~With the IC's guidance, the firefighters search each apartment for survivors and the source of the fire.~~ The IC is able to monitor the location of each firefighter and is aware of which apartments have been searched through the information provided on the GIS displays.

7. The EMS unit outside the apartment monitors the vital signs of all the firefighters in and around the fire scene. The unit alerts the IC that firefighter F725 is showing signs of distress, and the IC orders F725 and his partner F734 out of the building for a check-up with the EMS team.

8. Firefighter F765 pushes his emergency button when he becomes disoriented in the smoke. The IC immediately directs firefighter F788 to his aid by providing F765's location relative to F788 via three-dimensional geolocation information and the floor plan.

9. While other ~~the~~ firefighters check ~~every apartment~~ other apartments for victims, the ~~main fire is discovered in a second floor apartment kitchen where an electric range is burning.~~ fire in apartment 1202 is extinguished by E7 and L7 crew. Two adults and two children are discovered in the apartment suffering from smoke inhalation. They are carried outside the building where the EMS unit is ready to take over medical aid. RFIDs are attached to their arms and each is given an oxygen tank and mask to help their breathing.

10. While the firefighters put out the fire in apartment 1202, the IC checks the GIS display, which shows locations of ~~where the~~ fire personnel ~~are~~ and ~~where~~ all ~~the survivors and~~ rescued individuals ~~live in the apartment building.~~ ~~Two top floor~~ Several apartments on the 12th floor have not been searched, and the IC moves fire personnel to those apartments. The ~~apartment~~ building database indicates an invalid may be living in apartment ~~321.~~ 1208. The firefighters break down the ~~doors of both apartments and in 321~~ door of apartment 1208 and find a bedridden individual, who is in good condition, and a pet dog ~~in the other~~ another apartment. Both are taken from the building and outfitted with RFID devices.

11. ~~The fire is brought under control.~~ With the fire extinguished, the ~~The~~ IC releases E12, and the IC reconfigures E12's PSCDs for return to the fire station. E7 and L7 wrap up their fire operations, and A34 has to transport one fire victim to the hospital. The IC releases all remaining equipment and gives control to dispatch.

3.3.3 Fire Communications Summary

Throughout the scenario, the Building Information Server is collecting and transmitting to subscribers all real-time building sensor data which can then be displayed at the PSAP, or on the IC's mobile data computer. In addition, the building networks are aiding radio communications by acting as an extension of the IAN, bringing it into the building, and making a path for sending out voice, video, and health information to the responding fire (IC) and EMS personnel. The fire personnel and equipment, EMS support personnel, and the fire victims are also tracked by the network, providing geolocation information in real time, giving the IC and any other authorized personnel with current accountability of public safety personnel and of the fire's victims. All victim information and vitals are recorded through wireless monitors and voice recognition systems with no reliance on paper reports and notes. All fire personnel and equipment have monitors to measure vital conditions and status, which are reported by the PSCD and IAN network to the IC's GIS. The GIS also has access to city building department databases, which are searched and queried for building information and plans, fire hydrant locations, etc.

Appendix A.2 SoR v1.1 Section 4 Operational Requirements of PSWC&I, new Building Data Communications tables

4.2.6.5 Building Data Communications

Table XX: Fire Data Communications—Interactive 5

	The communication occurs:
with whom	Incident Commanders and other authorized users interact with the Building Information Server, which serves as interface to building control systems such as fire alarm system, smoke control, elevator control, and lighting system control.
for what purpose	To determine conditions within the building. To access building sensor information to be used in modeling tools to determine the most likely future course of events, such as predicted fire development. To enable control of building systems (such as smoke control or elevator mode) from external to the building as needed for incident mitigation.
with what special constraints	This sensor and building data must be current and accurate. Real time building sensor data from the building will add a real time element to Fire Department Pre Plans which are now simply static pieces of information.

4.2.7.4 Building Communications

Table XX: Fire Data Communication—Non-Interactive 5

	The communication occurs:
with whom	Upon alarm, the Building Information Server makes building information available to authorized users outside of the building. Information will be transmitted to Mobile Data Computers located in fire apparatus, IC vehicles, 9-1-1 centers, emergency operations centers, and other locations deemed appropriate.
for what purpose	Fire officers will have access to building information while responding to the incidents. An "En-Route" screen will give a plan view of the building with locations of fire hydrants, exterior fire department connections, approximate location of the incident, best access entrance and stairway to use, etc. An "On-Site" screen will have detailed information regarding the interior of the building with floor plans for each floor, exact location of the incident, type of alarm, status of elevators, location of interior fire department connections, location of building occupants, etc.
with what special constraints	This sensor and building data must be current and accurate. Real time building sensor data from the building will add a real time element to Fire Department Pre Plans which are now simply static pieces of information.

4.4.6.4 Building Communications

Table XX: Law Enforcement Data Communication—Interactive 4

The communication occurs:	
with whom	Incident Commanders and other authorized users interact with building control systems.
for what purpose	To determine conditions within the building. To access building sensor information to locate and track possible unauthorized individuals within the building. To communicate with building occupants to give them specific instructions during an emergency, lock and unlock doors from outside the building, monitor the location of elevators, etc.
with what special constraints	This sensor and building data must be current and accurate to provide situational awareness to Police Department building incident response. Some of these communications require encryption (e.g., door unlock commands).

4.4.7.4 Building Communications

Table XX: Law Enforcement Data Communication—Non-Interactive 4

The communication occurs:	
with whom	Upon alarm, the Building Information Server makes building information available to authorized users outside of the building. Information will be transmitted to Mobile Data Computers located in police cars, IC vehicles, 9-1-1 centers, emergency operations centers, and other locations deemed appropriate.
for what purpose	Police officers will have access to building information while responding to the incidents. An "En-Route" screen will give a plan view of the building with location of entrances to the building, approximate location of the incident, best building access to use, etc. An "On-Site" screen will have detailed information regarding the interior of the building with floor plans for each floor, exact location of the incident, type of alarm, status of elevators, location of building occupants, etc.
with what special constraints	This sensor and building data must be current and accurate. Real time building sensor data from the building will add a real time element to Police Department Pre Plans which are now simply static pieces of information.

Appendix A.3 SoR v1.1 Section 5 System of Systems, revised including building in network architecture

5 System of Systems

This section describes, in detail, the network topology that will be used in meeting the requirements set forth in sections 6, 7, and 8. Specifically, this section defines the network interfaces, both wired and wireless, and defines the links between the interfaces.

5.1 Network Description

The communications systems must be integrated with the public safety user's operations. For example, as a police officer leaves a patrol car to respond to a traffic stop or to investigate a domestic dispute, the critical communications capabilities, whether voice or data, must remain with the officer. As a firefighter enters a burning building, the biometric monitoring devices, the equipment status devices, and the firefighter's location device must indicate to the IC the firefighter's status and location at all times. These wireless devices must work in a variety of networks. Together, they will form the system of systems, with the following natural network hierarchy (see Fig. 1).

a. PAN—The PAN for a first responder can take on many different forms. Primarily, it is intended to
represent a set of devices on the person of a first responder that communicate with the first responder's PSCD as necessary. The devices on a PAN will include such items as heart rate monitors, location sensors, etc. This information could, and would in many cases, be transmitted to other areas of the network.

b. IAN—An IAN is a network created for a specific incident. This network is temporary in nature and is typically centered on a wireless access point attached to the first responders' vehicle. Multiple vehicles therefore dictate multiple wireless access points, all of which coordinate their coverage and transmissions seamlessly and automatically. For building incidents, the IAN may be extended by wireless access points inside the building(s). This network scales to the size of the incident, from a local traffic stop, to a large-scale, multi-discipline, multi-jurisdiction event.

c. JAN—The JAN is the main communications network for first responders. It handles any IAN traffic that needs access to the general network and provides the connectivity to the EAN. Additionally, it is the component of the network that will handle any and all communications from a first responder PSCD should a connection with the local IAN fail or be otherwise unavailable.

d. EAN—The local systems are, in turn, linked with county, regional, state, and national systems or EANs. It is expected that this network could be both wired and wireless, depending on the type of infrastructure deployed in the area, i.e., microwave point-to-point, fiber, etc.

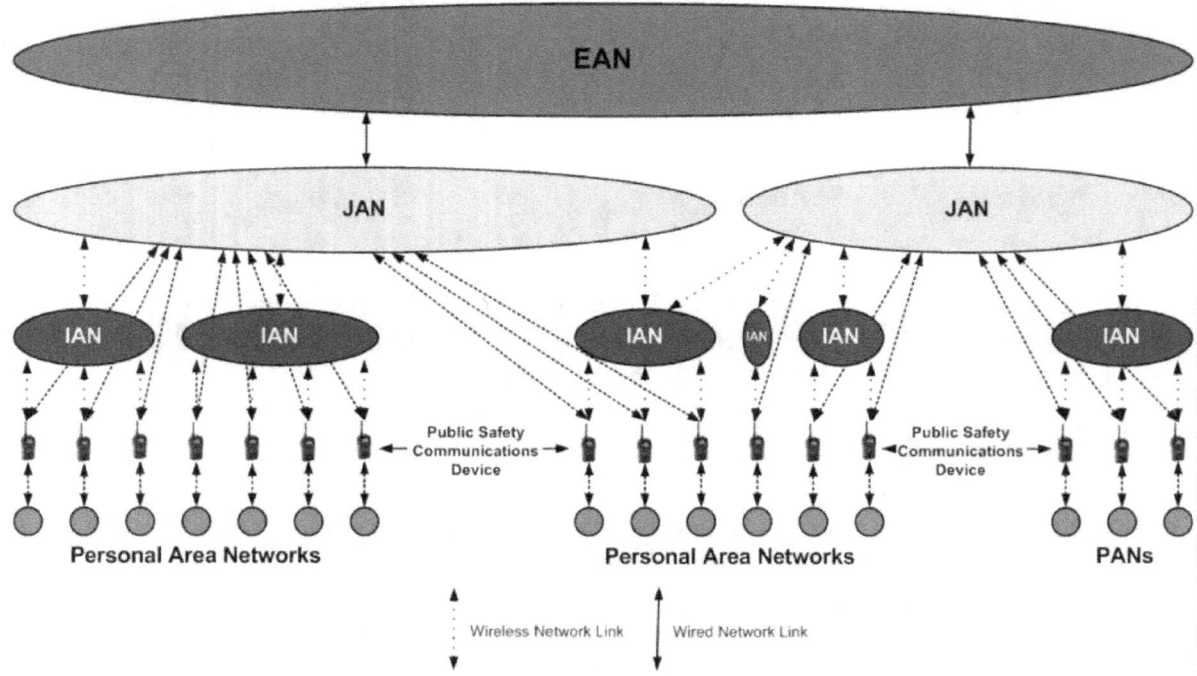

Figure 1:   Natural Network Hierarchy

Because public safety operations are usually conducted in the field and emergency operations must take place in the vicinity of the emergency, the networks must allow for mobile members, and/or the networks themselves must be mobile and temporary in nature. They must be dynamic and scalable to allow new resources to come onto a temporary network, and they must allow temporary networks to integrate with larger temporary or fixed networks.

Additionally, the management of these networks must allow for automated management as well as user led management, when necessary and as local policy dictates.

5.2 Network Diagram

The following network diagram shows all of the links and interfaces that have been identified based on the scenarios and requirements discussed in this document.

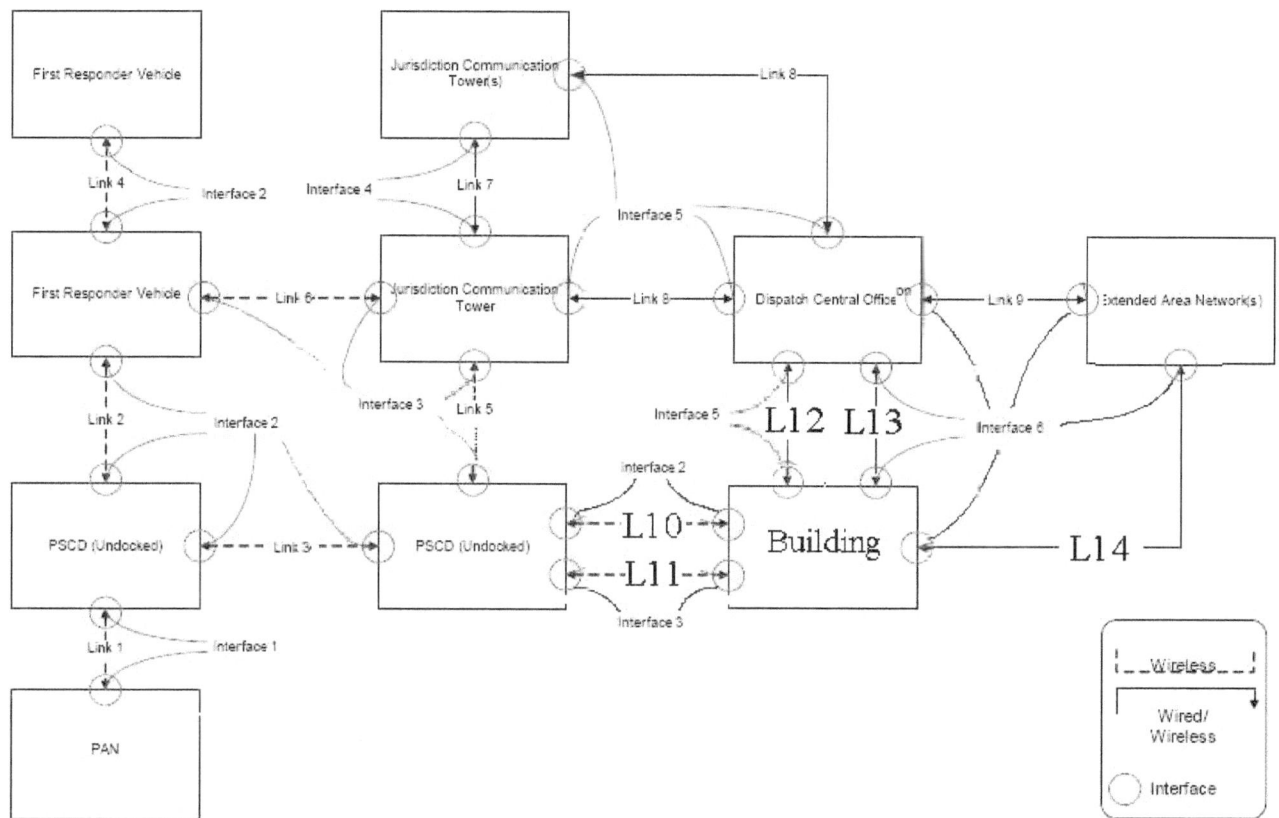

Figure 2: Link Diagram

In the preceding figure, the dotted lines denote a wireless connection, and the solid lines denote a wired connection. A red circle around the end of a link denotes a distinct interface, whether wired or wireless.

In the future vision of building incident management, the building itself has infrastructure in place to support building incident response communications. This can be in the form of providing in-building radio coverage, connectivity of responders inside to those outside, connection of the building site IAN to the JAN, as well as providing building situational awareness to incident commanders via data connections to intelligent building systems. This kind of in-building communications and data system will be of great service to first responders for large buildings with many occupants and complex systems. The building (as seen in Fig. 2) has multiple links and interfaces because it serves in multiple modes as follows:

Table XX

Mode	Name	Description
1	IAN member	Extends the IAN into the building via a distributed antenna system or via a connection across the building networks. In this mode the building acts like a "First Responder Vehicle" in Figure 2, but serves to establish connectivity between first responders outside and inside the building. This mode corresponds to Link 10.
2	Tower	The building uses an internal distributed antenna system to serve as a tower connection to the JAN for first responders inside the building who have no radio signal to the outside. This mode corresponds to Links 11 and 12.
3	Hybrid	Combines modes 1 and 2. The building extends the IAN into the building while offering a high-availability path to move data from the IAN to the JAN across a wired (or wireless) connection. In this mode, the building can serve as an auxiliary tower in out-lying areas with poor radio reception to connect the IAN to the JAN.
4	JAN server	The Building Information Server communicates data about the real-time status of the building to authorized clients on the JAN and EAN. This corresponds to links 13 and 14.

The following table provides a short description of each link in Figure 2.

Table 29: Network Diagram Link ~~Interface~~ Descriptions

Interface	Is Specified As
Link ~~Interface~~ 1	The PAN of the officer and the officer's PSCD. The data collected by the sensors on the officer's body is transmitted aggregately to the PSCD. The main considerations in sizing this link are the amount of data to be transmitted, the distance the transmission must travel, and interference from outside sources, including other officers' PANs.
Link ~~Interface~~ 2	The officer's PSCD and the first responder vehicle when the PSCD is in range of the vehicle's network, creating the IAN. Links 2, 3, 4, and 10 all use the same interface, but ~~separation~~ separating the links allows for separate performance specifications for each.
Link ~~Interface~~ 3	PSCD to PSCD communications. This link is used in peer-to-peer communications. Links 2, 3, 4, and 10 all use the same interface, but separating ~~separation~~ the links allows for separate performance specifications for each.
Link ~~Interface~~ 4	First Responder Vehicle to First Responder Vehicle communications. This link is used in vehicle to vehicle communications. Links 2, 3, 4, and 10 all use the same interface, but separating ~~separation~~ the links allows for separate performance specifications for each.
Link ~~Interface~~ 5	The officer's PSCD and JAN infrastructure. This connection is only used when the officer is out of range of the IAN created by the first responder vehicle, i.e., an officer on foot patrol would use this link fulltime while an officer operating out of a vehicle would only use this link when the

		connection between the officer and the vehicle is unavailable.
	Link ~~Interface~~ 6	The first responder vehicle and JAN infrastructure. This link is used for the same traffic that link 5 is used for, the primary differentiator being the location of the officer with respect to the vehicle. As was described for link 5, if the officer is within transmission distance of the vehicle, data is passed through the vehicle before being transmitted to the JAN, if not, the officer's PSCD transmits directly to the JAN.
	Link ~~Interface~~ 7	The JAN infrastructure pieces. While this connection is most likely also connected to the dispatch central office, it provides the capability to describe traffic that does not route itself through the dispatch central office.
	Link ~~Interface~~ 8	The dispatch central office and JAN infrastructure.
	Link ~~Interface~~ 9	The dispatch central office and a wider network. It is through this link that DMV, NCIC, PSTN, and other extranet queries will be forwarded.
	Link 10	PSCD to building. This link is similar to Link 2, and extends the IAN into the building via a distributed antenna or distributed access points to a wired network.
	Link 11	PSCD to JAN infrastructure. This link connects the officer's PSCD to the building acting as an auxiliary tower and would be used only when the IAN is not available. This may be the case if the internal building communications infrastructure connects to the JAN but not to the external IAN while the building structure blocks radio connection between the outside and inside.
	Link 12	Building to dispatch. When the IAN is not accessible inside the building, the building can act as an auxiliary tower to communicate with dispatch.
	Link 13	Building to dispatch and other authorized users on the JAN. The Building Information Server communicates data about the real-time status of the building.
	Link 14	Building to wider network. The Building Information Server communicates data about the real-time status of the building to authorized clients on the EAN.

Each of the interfaces identified in Figure 2 is unique to the network. The following table provides a short description for each interface.

Table 30: Network Diagram Interface Descriptions

Interface	Is Specified As
Interface 1	The interface that handles the aggregate transmissions to/from a first responders' PAN to/from the PSCD. This is a wireless interface.
Interface 2	The interface that handles transmission to/from the PSCD/First Responder Vehicle/ Building ~~to/from~~ from/to the PSCD/First Responder Vehicle/ Building via the IAN. This is a wireless interface.
Interface 3	The interface that handles transmissions to/from the PSCD/First Responder Vehicle/ Building ~~to/from~~ from/to fixed/mobile infrastructure

	via the JAN. This is a wireless interface.
Interface 4	The interface on a piece of fixed/mobile infrastructure that handles transmissions to/from another piece of fixed/mobile infrastructure. This interface can be wired or wireless.
Interface 5	The interface that handles transmissions to/from a piece of fixed/mobile infrastructure to/from the local dispatch central office. This interface can be wired or wireless.
Interface 6	The interface to/from other network types, including the PSTN, other jurisdictions, the public Internet, etc. to/from the dispatch central office. This is a wired interface.

Appendix A.4 SoR v1.1 Section 7 Public Safety Communications Device Functional Requirements, new "Building Communication System Functional Requirements" subsection

Section 7.4 Building Communication System Functional Requirements

The building acts as a communication device providing services to public safety officers in the event of a building incident. The building provides radio connectivity to connect PAN to IAN. The building also provides real-time information about the incident via an information server connected to the JAN or EAN. The Building Information Server (BIS) is a gateway device that gathers building automation and control system data and makes it available on the public safety network.

Matrix 35: Building Communication System Requirements

SoR Section 7.4 Requirement #	Qualitative Requirement Description	Additional Information
1	The building has a Building Information Server (BIS) that acts as a gateway to the public safety network.	See Matrix 36 for details
2	The building provides radio connectivity for responders	This could be via distributed antenna system with or without amplifier, or some other method.
3	Acts as an extension of IAN to link IAN to responder PANs inside the building.	The building provides network access points in the building with wired connections to the exterior. It acts in a role similar to a vehicle relaying IAN communications. It can also have an IAN-JAN link.

Matrix 36: Building Information Server (BIS) Requirements

SoR Section 7.4 Requirement #	Qualitative Requirement Description	Additional Information
1	Has BACnet interface to building control system to gather data from building sensors	BACnet (www.bacnet.org) is an ANSI and ISO standard for building control system communication.
2	Building data stream is made available to all authorized users to inform users of building incident status.	
3	Has database that (1) contains current floor plans with all public safety related static data indicated on plans, and (2) relates all available building sensors to physical locations in building.	Want to be able to see locations of static items like stand pipe locations, as well as real-time sensor info tied to floor plan.
4	Able to transmit data according to standard format such that it can be presented on floor plan in PSCD display	Message formats have not yet been standardized.
5	Provides locations of first responders relative to floor plan	
6	Saves logs of: (1) all building sensor data collected by the BIS; (2) all public safety users (clients) connected to the BIS	
7	Uses secure data transmission on interface to public safety network	The interface will implement all requirements specified for an interface to the EAN.
8	Provides, at minimum, a fire system connection, to gather fire sensor data points	
9	Updates real-time sensor data values at the BIS within a reasonable refresh period	Sensor data is not useful if old, and must report current conditions within a reasonable time scale.

Appendix A.5 SoR v1.1 Section 8.2 Network Functional Requirements, Matrix 96, additional building interface requirements

Section 8.2 Matrix 96

SoR Section 8.2 Requirement #	Qualitative Requirement Description	Additional Information
• • •		
6	The network must support the capability to interface with a building information server (BIS).	The BIS provides a building model (floor plans if not 3D), with information about all building systems, including real-time system sensor data (such as fire alarms, elevator status, and sprinkler flow status), presented on the floor plan.

Appendix A.6 Applications for the Building Information Server to Public Safety Network Interface

Following is a list of building applications that support public safety operations, with descriptions of the services, and possible updates to SoR section 6 tables to represent the functional requirements for these applications.

1. **Critical Alerts**
    - **What**: Building information server (BIS) initiates warning signal on certain events, e.g.:
        - Room went to flashover
        - Structure near collapse
        - Trapped occupant
        - Man down
        - Sprinklers on
        - New fire location (or new fire alarm)
    - **Who**: IC, dispatch, emergency communications center
    - **Network**: IAN, JAN
    - **Class of Service**: 2 (interactive, mission critical transactions)
    - **SoR Updates**: 6.1.2.3 (and 6.1.3.3)
        - Add new Requirement:
            - **Qualitative Requirements Description**: The network must support a signaling protocol for communicating mission-critical incident area sensor alerts. A sensor or sensor network communication device must be capable of establishing presence on the network.
            - **Additional Information**: Networked sensors in a building may communicate via a central node such as the building information system (BIS) directly on the IAN. Some vehicles such as HAZMAT trucks might have a similar vehicle information system. The Common Alerting Protocol under consideration by the OASIS Emergency Management Technical Committee may be adaptable for use as a wrapper for communicating alerts beyond the IAN.
    - **Notes**:
        - Push technology.
        - This application can be provided by the BIS interface that serves other applications. In this case the alert is a class of sensor events that are identified as critical and worthy of higher priority class of service. A much wider slice of the public safety community may be interested in only receiving critical alerts.

- Critical alerts could come from non-building sensors such as field sensors, or hazmat truck sensors, etc. This argues for a standard warnings message protocol that could be used by BIS or Vehicle Info Sys, etc.
- Critical alerts could come directly from sensors establishing presence on IAN
- Presentation of critical alerts could be made via blinking indicators on building display, text message, audio voice, or other means.

Note: The next three applications all (also) use the BIS communication interface. From a communications perspective, all three have the same requirements—to serve up building information based on a client one-time request or subscription to certain data. The content of the messages will be complex and dependent on the client making the request and software agent providing the data as well as the current status of the building and complexity of the BIS. Some protocol will be needed to establish what information is available and how to access it.

The following SoR update for Class of Service 3 is suggested to meet the requirements of these three applications.

- **SoR Updates**: 6.1.2.4 (and 6.1.3.4)
  - Add new Requirement:
    - **Qualitative Requirements Description**: The network must support transactions with software agents.
    - **Additional Information**: Agents may include: information repositories such as building floor plans or hazmat vehicle content information; real-time data provider with sensor status and event messaging (e.g. fire sensor alerts) and sensor event logs; and decision support modeling tools (fire predictions, plume calculations, collapse predictions).

2. **Building Static Data**
   - **What**: Retrieve information on static systems in building such as: floor plans, fire system (and other subsystem) component locations, and phone number listing by room, all of which are a part of a building information model (BIM)
   - **Who**: IC, dispatch, others
   - **Network**: IAN, JAN
   - **Class of Service**: 3 (interactive transactions)
   - **SoR Updates**: 6.1.2.4 (6.1.3.4)
     - Above
   - **Notes**:
     - There would be a one time get request for building info model file (or part of it), with no subscription and auto-updating.

- This by itself could be under Class of Service 4 (short transactions, bulk data)
- End application would present data on building display.
- Presentation could be 2-D or 3-D.
- Locating First Responders: If something like a "You are Here" RFID tag is placed in the building and PAN reads this and relays reading to application on IAN, then that application uses the corresponding building model and static data info to locate responder relative to floor plan.

3. **Real-Time Building Status**
    - **What**: Getting sensor event updates, current status of all sensors in building, sensor status/value histories (state of each sensor since start of incident), and sensor events log file (log of sensor events since start of incident).
    - **Who**: IC, dispatch, others
    - **Network**: IAN, JAN
    - **Class of Service**: 3 (interactive transactions)
    - **SoR Updates**: 6.1.2.4 (6.1.3.4)
        - Above
    - **Notes**:
        - This application involves either a get request to a web server (for logs and current status), or a simple push message from a publish subscribe server after having subscribed.
        - End application could request data by category.
        - Presentation could be 2-D or 3-D.

4. **Decision Support**
    - **What**: Getting Sensor Driven Fire Model (SDFM) and other decision support modeling results
    - **Who**: IC, dispatch, others
    - **Network**: IAN, JAN
    - **Class of Service**: 3 (interactive transactions)
    - **SoR Updates**: 6.1.2.4 (6.1.3.4)
        - Above
    - **Notes**:
        - End application would present data modeling results on building display.
        - Presentation could be 2-D or 3-D. Could also be text or non-visual such as a voice giving directions.

5. **External Control:**
    - **What**: the IC can send commands back to BIS, allowing IC to: open doors, control elevator, direct HVAC smoke control, etc.
    - **Who**: IC
    - **Network**: IAN, JAN
    - **Class of Service**: 3 (interactive transactions)
    - **Notes**:
        - This application has been discussed, but we view it as a longer range goal. It is not clear now what would be useful for the public safety community. When would someone offsite (such as on the way, or at dispatch) need to control building systems? Once onsite the IC has ability to control systems already.
        - Many more issues with this: security, command and control, safety, difficulty of translating command and reversing data path.
    - **SoR Updates**: 6.1.2.4 (6.1.3.4)
        - None suggested at this time.

- SoR service classes:
    - Class of Service 0: mission critical, jitter-sensitivity, high interaction
    - Class of Service 1: same, but not mission critical
    - Class of Service 2: highly interactive transaction data
    - Class of Service 3: less critical interactive transactions—this would be a good class for sensor updates
    - Class of Service 4: low-loss traffic: short transactions, bulk data.

- Keep building network traffic separate. Voice and video data from PAN may traverse building nets, but that enters and exits building network via standard IAN interfaces. The building appears as if it is a vehicle or group of vehicles.

SoR input, Section 4 tables:

SoR Section 8.2 update

This section covers the non-public safety networks that will need an interface for sending and/or receiving information from public safety. All of these interfaces will be made through the EAN.

Matrix 96: Interfaces Requirements

SoR Section 8.2 Requirement #	Qualitative Requirement Description	Additional Information
1	The network must support the capability to interface with the PSTN.	
2	The network must support interfacing with public utility information, such as that for the power grid, natural gas distribution systems, etc.	The level of security needed for this interface needs to be understandably high due to concerns of terrorist attacks on such utilities.
3	The network must support the capability to interface with non-public safety data networks, including the Internet, in a secure manner.	An example service is the IamAlive service. Other examples include communications with Public Health agencies, Emergency Management Departments, and other pertinent organizations.
4	The network must be capable of accessing real-time weather information.	Weather information can take the form of a forecast, current weather at a given site, network sensors, etc.
5	The network must support the capability to interface with the Department of Transportation's Intelligent Transportation System (ITS).	
6	The network must support the capability to interface with a building information server.	A standard building information server interface (building side) is still under development. The public safety network should be able to access real-time incident information from the building server.

Appendix B: NIST SoR Questionnaire Feedback

F-1 Response of Richard Elliott, Gaithersburg, MD, Police Department, to NIST's "Questions for Public Safety Practitioners r.e. the SAFECOM Statement of Requirements and the addition of the building component"

Aug 9, 2006

1. What do you think about the Statement of Requirements?
    a. do the stated requirements make sense? **Yes**
    b. do you see any missing requirements? **No**
    c. do you believe any of the requirements are "over-the-top" and not necessary? **No, but a good question that often does not get asked**

2. Does your fire district have any code requiring building owners to provide radio reception inside large buildings?
Unable to answer, more geared to FD or Fire Marshal

3. Do you know of any building system technology or installation that gives the incident commander situational awareness beyond what can be learned from looking at the fire panel? **At present the only thing I have run into that would qualify is CCTV, of course in that case it would be necessary for the IC or designee to physically go to the monitor room, a situation not always possible given fire or other tactical situations. I am unaware of any private buildings that have the capacity to stream video to an outside source. There are probably sensitive government buildings where that is possible, but I could not name any.**

4. Does the role of the building, as presented in Section 5 and Section 7.4 make sense and seem appropriate? **It does**

5. Would the section 5 figure 2 Link Diagram work better if the "Building" were replaced with a more generic "external sensor network" that could be a building or vehicle or outdoor special area with sensors and wireless networking? **The change would be insignificant, external sensor network is a more "catch all" phrase and would probably cause less change to be made in the future as the technology advances to a point where outdoor and vehicle monitoring would be more feasible, both from a technological and economic standpoint.**

6. Do you think that Fire, Police, or Emergency Medical Services would benefit by the ability to obtain real time sensor information from various sensors within or outside of a building? **This one is a "no brainer" emergency services would definitely benefit from such information. At present information can go through several sources before it gets to the end user (sensor to alarm company to 911 call taker to dispatcher to IC to field unit). From my experience, this can not only take enough time to render the information useless, but at each stage the possibility**

exists for changes that may render the information completely different. By eliminating the numerous middle men, the time factor becomes less critical and the information is more likely to be accurate.

7. Do you think that Fire, Police, or Emergency Medical Services would benefit by requiring building owners to provide an extension of the Incident Area Network?
Again this one is a yes, but of course there are political considerations involved as it appears this in not an inexpensive proposition. The average business owner would probably balk at the installation and maintenance expense being mandated. The more sensitive or high value the building, the more likely the owners would be amenable to installation.

8. Do you think that Fire, Police, or Emergency Medical Services would benefit by requiring building owners to provide a building information server that might require frequent maintenance by owner and inspection and testing by fire dept?
Again yes, but the circumstances listed in answer 8 would still be an issue.

F-1 Response of Chief Keith Richter, Contra Costa County (CA) Fire Department, to NIST's "Questions for Public Safety Practitioners r.e. the SAFECOM Statement of Requirements and the addition of the building component"

Aug 9, 2006

1. What do you think about the Statement of Requirements?
    a. do the stated requirements make sense? **yes**
    b. do you see any missing requirements? **no**
    c. do you believe any of the requirements are "over-the-top" and not necessary? **Requirements need to be based on the amount of risk associated with a particular building or occupancy. The cost of a system needs to be weighed against the increased benefit of a system.**

2. Does your fire district have any code requiring building owners to provide radio reception inside large buildings? **No**

3. Do you know of any building system technology or installation that gives the incident commander situational awareness beyond what can be learned from looking at the fire panel? **No**

4. Does the role of the building, as presented in Section 5 and Section 7.4 make sense and seem appropriate? **Yes**

5. Would the section 5 figure 2 Link Diagram work better if the "Building" were replaced with a more generic "external sensor network" that could be a building or

vehicle or outdoor special area with sensors and wireless networking?
I think the graph is clear, perhaps a footnote stating that "Building " could be any type of network.

6. Do you think that Fire, Police, or Emergency Medical Services would benefit by the ability to obtain real time sensor information from various sensors within or outside of a building? **Clearly, yes.**

7. Do you think that Fire, Police, or Emergency Medical Services would benefit by requiring building owners to provide an extension of the Incident Area Network?
I would say that the larger the building or area, the more benefit derived from the IAN extension.

8. Do you think that Fire, Police, or Emergency Medical Services would benefit by requiring building owners to provide a building information server that might require frequent maintenance by owner and inspection and testing by fire dept?
Again, I would say that a threshold for minimum size would need to be established to make the cost/benefit analysis justify the investment.

Additional Feedback on SoR
The document (SoR v 1.1 with NIST BFRL revisions) took me a while to digest and it was very educational for me. Overall, it's obvious that a lot of work has already been done to assemble this book. I wish I had more technical knowledge about the communications aspect, but I know there are other groups covering that. I have a few observations to offer:

Article 3.3.2 #6
The last sentence talks about GIS tracking of firefighters. You might want to address how the system would track and record searched apartments. Perhaps the voice recognition discussed elsewhere could also log the primary/secondary search function.

Same article #8
Activation of emergency button would be more effective if it alerted all personnel and gave them coordinates to the distressed firefighter. The IC could then coordinate the rescue effort and avoid the "candle-moth scenario" when firefighters are down.

Article 4.2.6.5
Table marked "XX". Under "for what purpose" block there is a phrase "such as predicted fire development". It seems that building sensors could also indicate structural integrity or instability as in the case of earthquake damage or prolonged heat exposure to structural members.

Article 4.2.7.4
Table marked "XX". Under "for what purpose" block there is a phrase

"status of elevators". I think it would be more descriptive to say "status of building's internal systems". This could be elevators, alarm systems, HVAC, water flow, etc.

Table 29-Network Diagram Link Descriptions
I'm not sure exactly where to insert this idea, but I think it might go with the definition of Link 9 (between a central dispatch and a wider network). This talks about extranet queries such as DMV, NCIC, etc. A project I'm currently working on in CA is to track mutual aid resources within the coastal region using a web-based secure portal. Once launched, this type of database could be queried by a dispatch center to determine availability of additional resources if needed. This capability could be applied across disciplines for any type of mutual aid. One difficulty currently is that there is no mandated P25 equivalent for CAD systems. Interface 6 on Table 30 would be an example of where a CAD to CAD interface could export resource status into a mutual aid database that could be posted on the web. I think this would be a great enhancement to the primarily manual methods currently used to track mutual aid resources.

Article 6.2.1 Authentication
Mutual aid resources would also be more easily authenticated if there were an interface that allowed CAD to CAD communication. Could a major incident be scaled up in size by interfacing CAD and RF systems? Could GPS tracking be exported from the home agency's CAD to the receiving agency's CAD? It's beyond my level of understanding, but some smart guys might figure it out.

Section 7.4 Matrix 36 (typos)
#1 Far right column, "BACnet is an" (not "and").
#9 middle column, "perios" should be "period".

C.2.1 Scenario
#9 "Dispatch" should be "First unit".

C.2.2 Narrative
#13 I suggest changing "Stang this thing" to "set up master stream operations". A stang, I believe, is a manufacturer's name that has become a slang term for a monitor.

Appendix C: Standards for Building Information Exchange with First Responders: presentation to the Alarm Industry Coordinating Council

Presented June 12, 2008 by David Holmberg

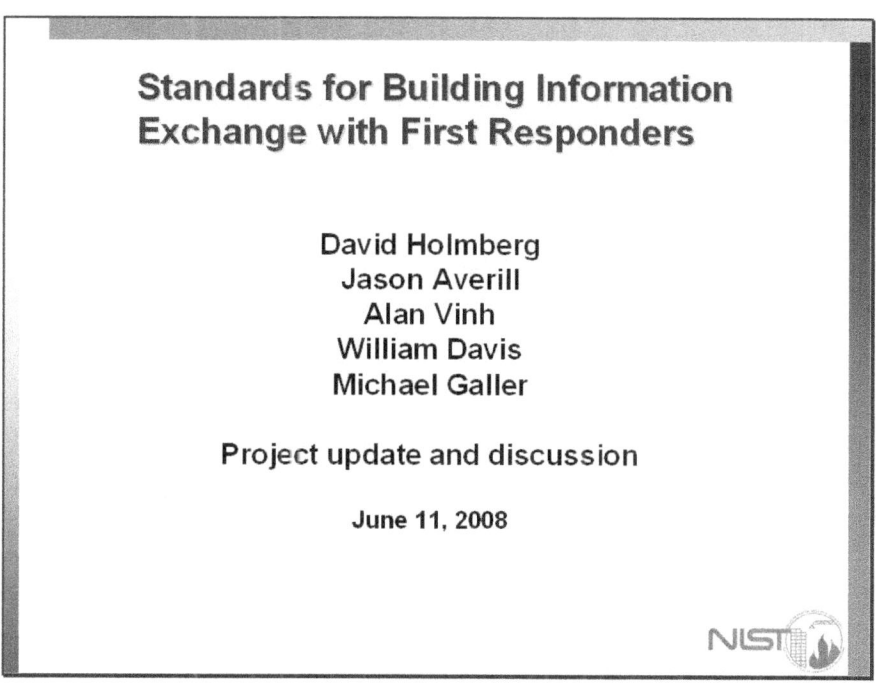

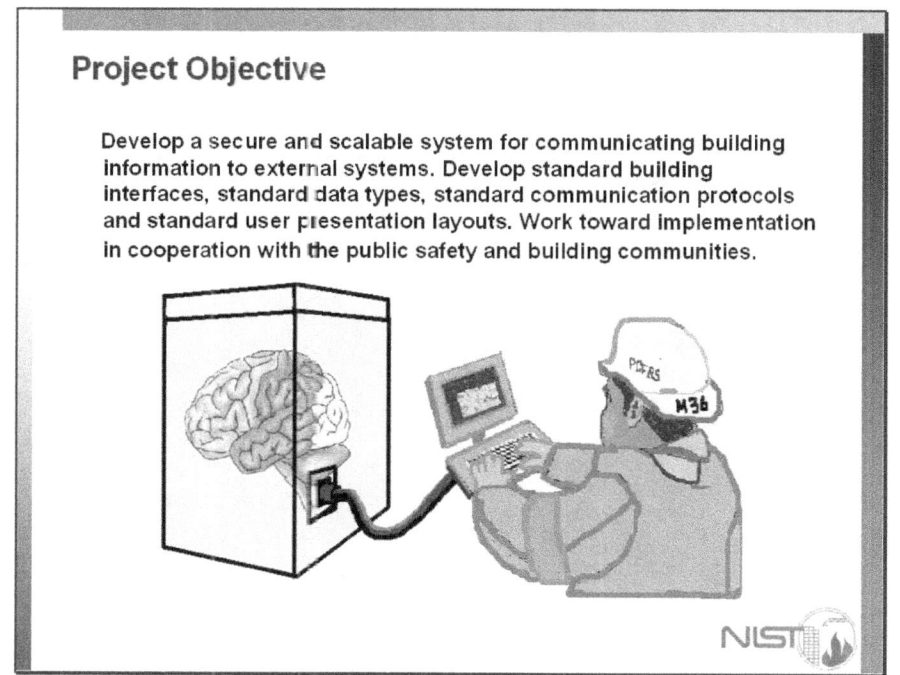

BIS Interface Stakeholders

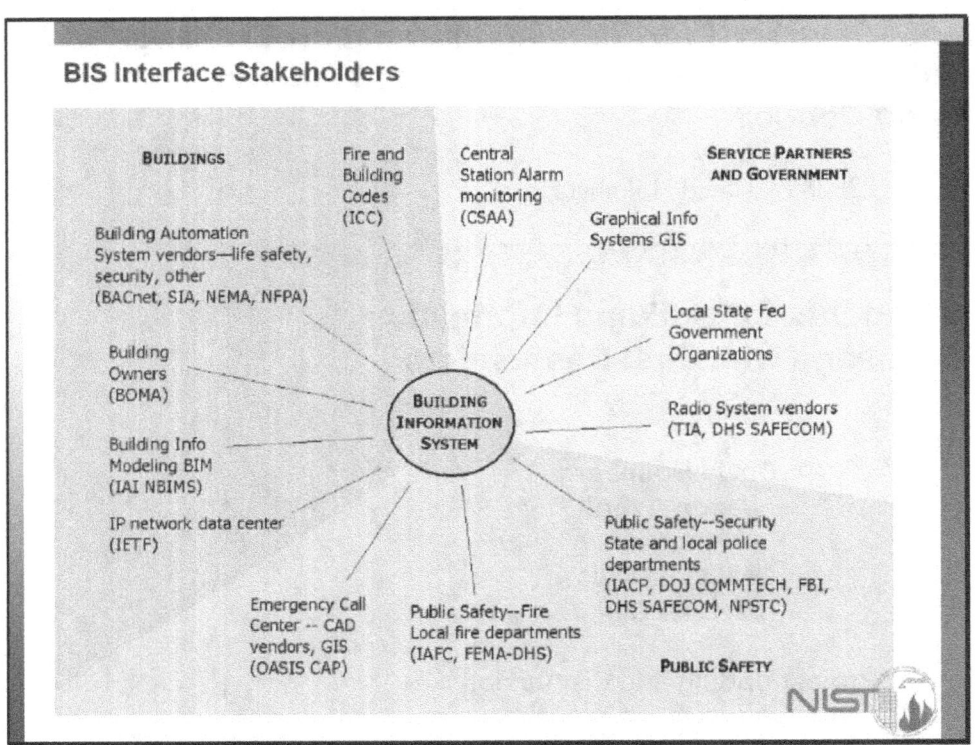

What are the technical challenges?

- A building information system that can handle generic requests ("where's the fire?"), abstracting control system specifics.
- Scalable network of servers to support secure delivery of alerts and data exchange with public safety users.
  - Detailed information for an Incident Commander
  - Filtered alerts for public safety officials monitoring an incident
- Data model for extracting information from building and presenting in understandable format to public safety.
- Providing limited outside control of building systems.
- Integrating building sensor data with stationary building data (floor plans, utility access points, fire and security system details, etc.)
- Integrating public safety sensor networks with fixed building networks.

What NIST is doing about this problem

- Developing the Building Information Services and Control System (BISACS) as a secure and scalable prototype system to communicate building information to external systems
- Ongoing effort to develop standard building interfaces to give authorized public safety users access to building intelligence
- Ongoing effort to develop standard user interfaces, data encapsulation standards and communication methodology standards for presenting information to the first responders (SB30 Building Interfaces Task Group)
- Working with NENA to look at how to move building alerts on to the NG9-1-1 network.

Building Information Services And Control System (BISACS)

- The BISACS is a network of computers designed to monitor and control devices from small to very large areas.
- Alerts are sent from these computers up the network hierarchy to human monitoring stations where this information can be acted upon accordingly.
- The data communicated between these computers are encrypted and secured.
- User authentication and authorization is required for accessing the BISACS network.
- Future plans include the ability to remotely send commands to buildings such as commands to shut off utilities.

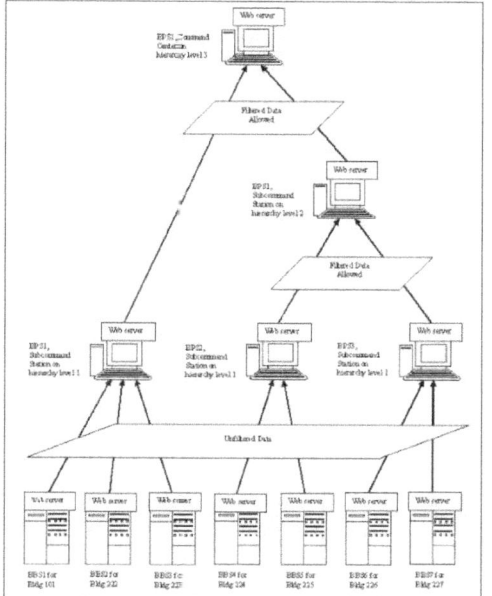

The Common Alerting Protocol (CAP)

- Alerts sent between all computers within the BISACS network are encapsulated using the Common Alerting Protocol version 1.1 standardized by The Organization for the Advancement of Structured Information Standards (OASIS).
- The Common Alerting Protocol is a simple but general format for exchanging all-hazard emergency alerts and public warnings.

The Common Alerting Protocol (CAP)

```xml
<?xml version="1.0" encoding="UTF-8" ?>
<alert xmlns="urn:oasis:names:tc:emergency:cap:1.1">
    <identifier>1179353147004</identifier>
    <sender>https://p623572.campus.nist.gov:8443/bisacs</sender>
    <sent>2007-05-16T18:05:47-04:00</sent>
    <status>Exercise</status>
    <msgType>Alert</msgType>
    <source>alarm1bundle.sensor01</source>
    <scope>Public</scope>
    <info>
        <category>Fire</category>
        <event>Smoke</event>
        <urgency>Immediate</urgency>
        <severity>Extreme</severity>
        <certainty>Observed</certainty>
        <expires>2007-05-16T18:06:47-04:00</expires>
        <description>Smoke detector, building 226, 3rd floor, room B346.</description>
    </info>
</alert>
```

• This is a sample Common Alerting Protocol message

• Client applications will filter on data based on the content of these messages.

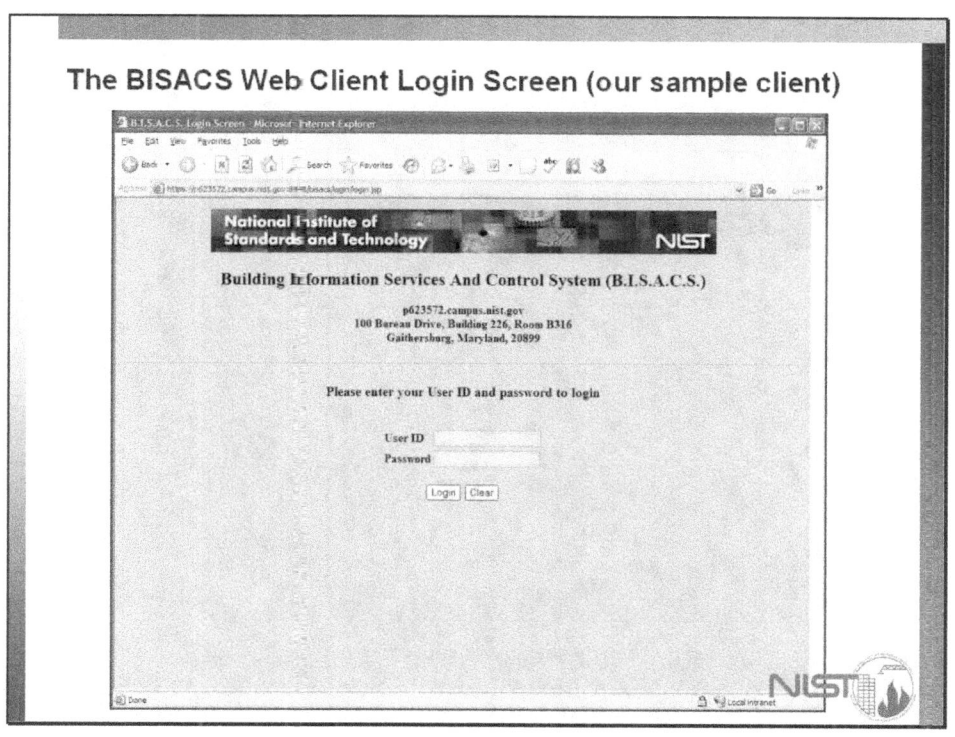

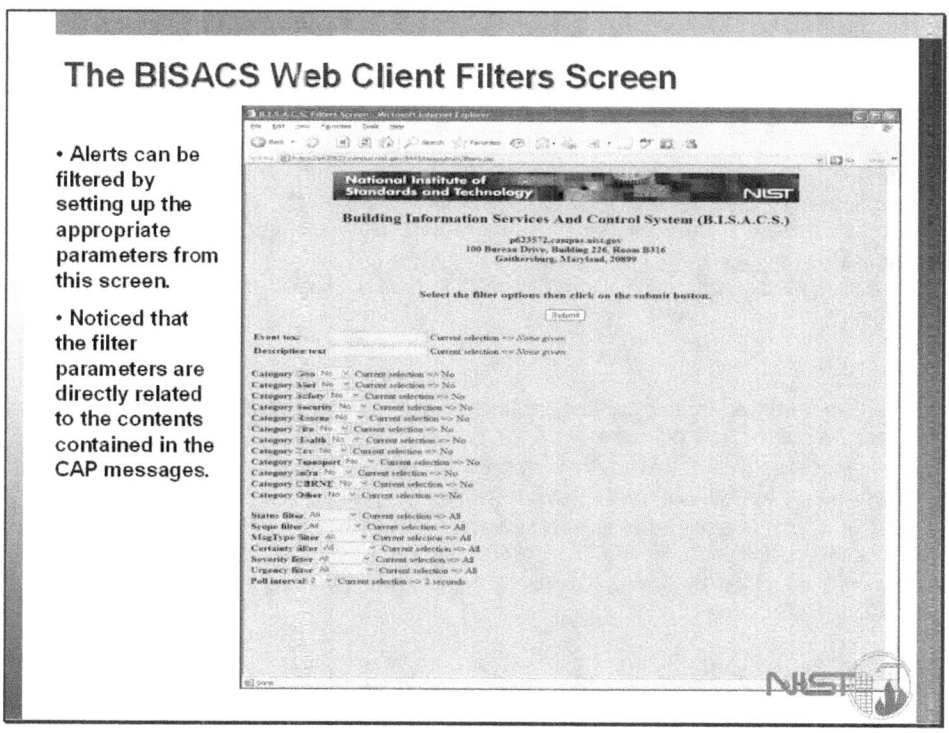

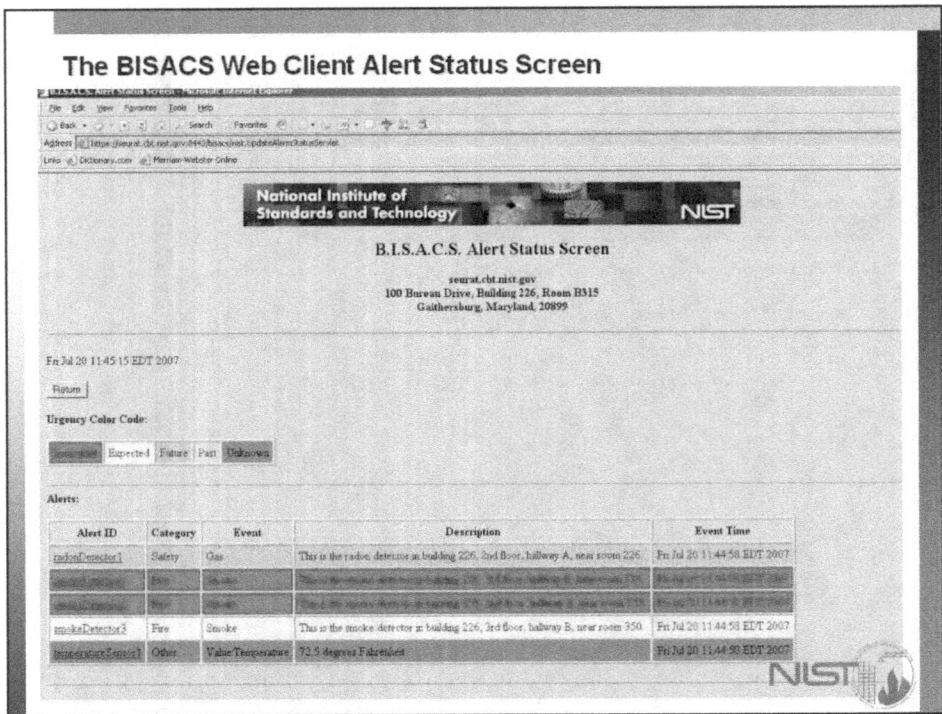

NG9-1-1 Overview

The Problem:
- The rapid evolution of the types of devices that can be used to call for help requires major changes in the existing emergency services architecture
- There is an increasing volume and diversity of information that can be made available to assist Public Safety Answering Points (PSAPs) and responders in an emergency

NG9-1-1 Vision:
- Create an all-IP-based emergency communications system that will run over an Emergency Services IP network (ESInet)
- ESInet is designed as an IP-based inter-network (network of networks) shared by all agencies which may be involved in any emergency
- The PSAP is capable of receiving IP-based signaling and media for delivery of emergency calls conformant to the NG9-1-1 interface standard
- Common Alerting Protocol Messages are routed to their proper destinations for processing

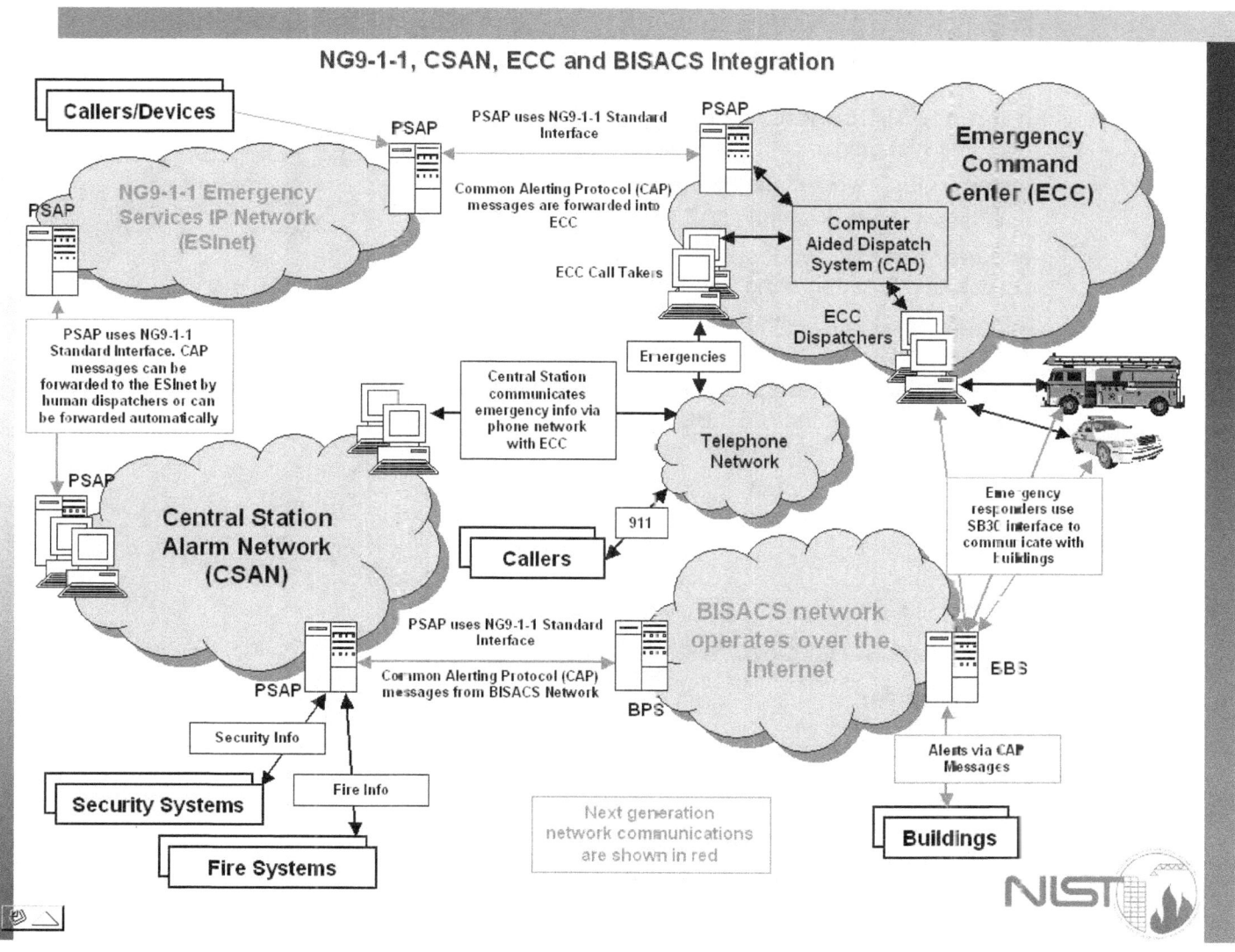

Road Map for System Implementation

- Install a NIST BBS/testbed on the Internet with live alerts (cybute1.nist.gov)
- Install an ECC work station to monitor the NIST BBS
- Implement the NSI on the NIST BBS
- Integrate/test ECC work station with the NIST NSI in preparation for integrating with the NG9-1-1 system
- Integrate NIST BBS with the NG9-1-1 system to send alerts through the ESInet
- Integrate/test ECC work station with the NG9-1-1 system to receive alerts from the NIST BBS. This validates end-to-end communication

Results so far

This project builds on and collects together some previous efforts. Up to this point we have made significant progress:

- Hosted workshops to identify building information needs of fire and police responders, and to find "what is working" for high-rise and complex incident response communications.
- Developed a user interface for presentation of building information to fire and police with demonstrations at NIST and in Wilson, NC.
- Produced a video with distribution count past 15,000
- NEMA SB-30 2005 standard for remote fire panel display.
- Investigated the potential for moving first responder communications over existing building networks.
- Worked with SAFECOM to address the role of buildings in the SAFECOM Statement of Requirements.
- Completed implementation of a first generation Building Information Services And Control System (BISACS) using building alerts encapsulated in the OASIS Common Alerting Protocol (CAP).

www.ingramcontent.com/pod-product-compliance
Lightning Source LLC
Chambersburg PA
CBHW080301180526
45167CB00006B/2626